WHY YOU CAN NEVER GET TO THE END OF THE RAINBOW

AND OTHER MOMENTS OF SCIENCE

WHY YOU CAN NEVER GET TO THE END OF THE RAINBOW

AND OTHER MOMENTS OF SCIENCE

EDITED BY   **DON GLASS**

SCIENCE ADVISOR,   **PAUL SINGH**

ORIGINAL SCRIPTS BY   **STEPHEN FENTRESS**

INDIANA UNIVERSITY PRESS   **BLOOMINGTON · INDIANAPOLIS**

These essays are adapted from scripts for the radio series *A Moment of Science,* produced in the studios of WFIU, Bloomington, Indiana.

Illustrations by Michael Cagle

The paper used in this publication meets the minimum requirements of American National Standard for Information Sciences—Permanence of Paper for Printed Library Materials, ANSI Z39.48-1984.

Manufactured in the United States of America

Library of Congress Cataloging-in-Publication Data

Why you can never get to the end of the rainbow and other moments of
   science / edited by Don Glass ; science advisor, Paul Singh ;
   original scripts by Stephen Fentress.
         p.   cm.
      Summary: Short essays explain such scientific questions as why
   cats' eyes glow at night, why rivers don't flow in a straight line,
   and how the world looks to a bee.
      ISBN 0-253-32591-9 (cloth : alk. paper). — ISBN 0-253-20780-0
   (pbk. : alk. paper)
      1. Science—Miscellanea—Juvenile literature.   2. Moment of
   science (Radio program)   [1. Science—Miscellanea.]   I. Glass,
   Don, 1939– . II. Fentress, Stephen.
   Q163.W497   1993
   500—dc20   92-34770

1 2 3 4 5 97 96 95 94 93

CONTENTS

PREFACE

The short essays contained in this book are adapted from the scripts for the radio series *A Moment of Science*. The series of two-minute radio programs was conceived by Paul Singh, professor of physics at Indiana University, and produced and distributed by WFIU, the university's public radio station. Dr. Singh served as the science producer, and Don Glass produced the radio programs.

The intent of the programs was to present the relevancy of scientific findings in a manner that was clear, concise, and unintimidating. Unnecessary jargon was avoided, and the material was presented so that it could be understood and appreciated without prior scientific knowledge.

Since its national debut in 1988, listeners to *A Moment of Science* have regularly asked for transcripts of the programs. It seemed that a logical step would be to prepare a selection for publication.

All of the original scripts in this volume were written by Stephen Fentress. He wrote each program so that it would be an enjoyable diversion which would move the listener to think, "Wow, that's neat!" and go tell somebody about it. We have removed some on-the-air business and updated a few items, but otherwise we present this material as it was broadcast.

We purposely have not grouped the scripts by scientific area, because we wanted to maintain that little element of surprise as the reader moves from one to the next. The short format of the radio programs was based on the premise that people have limitless curiosity, but limited time. That applies to this book as well: each selection is self-contained, so you can read one or a dozen at a time, depending on how much time you have.

If after reading these *Moments of Science* you have a little better understanding of the world around you, then our goal has been achieved.

Don Glass and Paul Singh
Bloomington, Indiana

ACKNOWLEDGMENTS

Without the cooperation of the administration and scientific community of Indiana University, neither the radio series nor this book would have been possible.

Funding, of course, is critical to any undertaking, and for that we thank the Dean's Incentive Fund of the College of Arts and Sciences, the Office of Research and Graduate Development, the Office of the Vice-President of the Bloomington campus as well as the Office of the Chancellor, the Office of the President of the University, and the Indiana University Radio and Television Services.

We express our sincere appreciation to the scientists who generously gave of their time by checking the copy for scientific accuracy. Those who helped with this series include Barry Aprison, Biology; Alan Backler, Teaching Resources Center; Tom Blumenthal, Biology; John Castellan, Psychology; James Craig, Psychology; David Easterling, Geography; Guy Emery, Physics; John Ewing, Mathematics; Daniel Knudsen, Geography; Harold Ogren, Physics; David Parkhurst, Public and Environmental Affairs; William Popkin, Law; Robert Port, Linguistics; Elizabeth Raff, Biology; Rudolf Raff, Biology; Steven Russo, Chemistry; and Eugene Weinberg, Medical Sciences and Biology.

WHY YOU CAN NEVER GET TO THE END OF THE RAINBOW

AND OTHER MOMENTS OF SCIENCE

Benjamin Franklin and the Swatches on the Snow

In a letter written in 1761, Benjamin Franklin tells how he collected some little squares of broadcloth, tailor's samples of different colors: black, dark blue, light blue, green, purple, red, yellow, and white. He wanted to demonstrate that these colors would absorb different amounts of light from the sun and convert the light to different amounts of heat.

On a bright winter day, when the ground was blanketed with freshly fallen snow, Franklin laid the cloth squares on the snow, in the sun, and left them for a few hours. When he came back, he saw that the black square had sunk deeper into the snow than any of the others. The dark blue cloth had sunk a little less, and the white square not at all. Each of the other squares had melted its way down to some in-between depth.

Even two hundred years ago, people knew that dark-colored things get warmer in the sun than light-colored things. But Franklin's little experiment demonstrated it scientifically by comparing cloth samples that were all the same except for one thing: color.

The dyes in the samples absorbed different amounts of sunlight. The black cloth got the warmest because it absorbed all the colors of the sunlight and reflected almost none—that's why it looked black. The white cloth stayed the coolest; it reflected all the colors of the sunlight and absorbed very little light. The red cloth absorbed some of the sunlight and warmed to a medium temperature, but it reflected red light to the eye and looked red. And so on for the other colors.

Benjamin Franklin did other experiments, especially with electricity, and even speculated about whether flies drowned in wine could be brought back to life. But that's another story.

N. G. Goodman, ed., *The Ingenious Dr. Franklin: Selected Scientific Letters of Benjamin Franklin* (1931; Philadelphia: University of Pennsylvania Press, 1974), p. 181.

Once You've Had It, You Can't Catch It Again

The human body fights infection by recognizing the invader. Suppose the invader is a virus. Viruses are a threat because they take over living cells and use the machinery inside to make more viruses. This wrecks the cells.

But before a virus reaches a cell, it may run into an antibody. Antibodies are giant molecules, shaped a little like twisted coat hangers. Your body has about a million different types of antibodies. An antibody fits its corresponding virus like a key in a lock.

When an antibody attaches to a virus, it may spoil the virus's ability to invade cells. Or the antibody may simply put a chemical tag on the virus that says, in effect, "White blood cells, please destroy me."

Once the invading virus is recognized, mass production of the corresponding antibody begins. It takes a few days, but if the body can make antibodies faster than the virus can make viruses, the viruses lose and the body gets well. The disease AIDS, by the way, represents a failure of this system.

After the body gets well, what's left? White blood cells, mopping up. Unused antibodies. And memory cells that save the recipe for the antibody just used. If the same virus ever returns, it will be defeated much faster the next time. That's why you don't get chicken pox again. "Howdy, viruses," your immune system says. "You're only a stranger here once."

N. K. Jerne, "The Immune System," *Scientific American*, July 1973.
Helena Curtis, *Biology*, 4th ed. (New York: Worth, 1983).

Why is the sky blue?

Why is the sky blue? It can't be that the atmosphere has a blue color like the blue of a tinted windshield. In that case, going outside in the daytime would be like walking around inside a blue glass bottle, with a blue sun shining blue light everywhere, and blue stars and a blue moon at night.

The blue can't be from dust, because the air over gravel parking lots and quarries is whitish, not bluish.

The blue can't be the result of water droplets. Clouds are made of water droplets, and clouds are white. It's not a matter of relative humidity, either; a dry sky over Arizona can be just as blue as a humid sky over Minnesota.

Blue is not the color of outer space. The background of space is black. So, the sky is black at night—it's blue only during the daytime, when the sun is shining on the atmosphere.

The sun shines with all the colors of the rainbow—blue, yellow, red, and all the rest—mixed together to make white light. The reds and yellows pass through air easily, but some of the blue portion of sunlight is scattered in every direction by air molecules. When you look to the sky on a clear day, you see blue light scattered from sunbeams by molecules of nitrogen, oxygen, and carbon dioxide.

The more air, the more scattering. In early morning and late afternoon, the sun's light passes through so much air that most of the blue has been scattered away by the time the light reaches you. The reds and yellows remain and the sun looks reddish.

Marcel Minnaert, *The Nature of Light and Colour in the Open Air* (New York: Dover, 1954).

Richard P. Feynman, *The Feynman Lectures on Physics* (Reading, Mass.: Addison-Wesley, 1964), vol. I, chap. 32.

STATIC ELECTRICITY

Static electricity reveals the nature of the so-called electrical force, one of the basic forces of the universe.

In some ways the electrical force is like the force of gravity. Electrical and gravitational forces get stronger if the interacting particles get closer, and weaker if they get farther apart. Actually, the electrical force is about a billion billion billion billion times stronger than gravity at the same distance.

Unlike gravity, which always pulls things together, the electrical force can pull particles together or push them apart, depending on the so-called charges of the particles. Electrical charges can be positive or negative. Like charges repel; opposites attract. Positive and negative charges attract, but two positives or two negatives repel each other. (Incidentally, some elementary particles, such as neutrons, have no electrical charge.)

Electricity holds atoms together. The nucleus at the center of every atom has a positive charge; the electrons surrounding the nucleus have negative charges, so they stay nearby. In most objects the positive and negative charges are exactly equal in number, so the net charge is zero.

Sometimes, though, everyday objects pick up a few extra electrons or lose a few electrons. When that happens, the object has a net static charge—static in the sense that those displaced electrons stay where they are, rather than flowing as a current. A charged object, like a bed sheet coming out of a clothes dryer, exerts noticeable electrical forces on other objects—like socks in the same load.

Richard P. Feynman, *The Feynman Lectures on Physics* (Reading, Mass.: Addison-Wesley, 1964), vol. II, chap. 1.

A. D. Moore, "Electrostatics," *Scientific American*, March 1972.

WHY ONE ROTTEN APPLE CAN SPOIL THE BARREL

If you buy green, underripe lemons or bananas, you can make them ripen faster by keeping them in a paper bag.

Chinese growers used to ripen fruit by keeping it in a room with burning incense.

In the West, farmers used to "cure" fruit with kerosene stoves.

What's behind all these processes is a gas called ethylene. Ethylene comes from burning fuels like kerosene, and it's made naturally by all parts of a plant at one time or another. Ethylene stimulates germination of the seed, flowering, ripening of fruit, dropping of fruit, and dropping of leaves. The ethylene produced by ripe fruit will stimulate another nearby fruit to ripen. So there's a chemical basis for the proverb about one rotten apple spoiling the whole barrel.

Ethylene is used in commercial agriculture to stimulate the ripening of bananas, tomatoes, and citrus fruits and to help give them the colors people expect.

Fruit shippers usually want to stop the ripening process, so they store fruit in rooms that have ethylene chemically removed from the air.

Incidentally, farmers at one time thought that it was the heat from a kerosene stove that was ripening their fruit. But those who tried modern non-kerosene heaters didn't get the results they wanted. It wasn't the heat that was "curing" the fruit—it was the ethylene.

"Ethylene," in *McGraw-Hill Encyclopedia of Science and Technology* (New York: McGraw-Hill, 1987).

Frank B. Salisbury and Cleon W. Ross, *Plant Physiology*, 3rd ed. (Belmont, Calif.: Wadsworth, 1985).

SUPERCONDUCTORS

Superconductors are materials that are radically different from ordinary materials. Superconductors conduct electric current with absolutely no resistance. Ordinary materials, including the metals we use to make wires in electrical appliances, offer some resistance to electric current. That's why electrical appliances get warm.

Thousands of metals and alloys and other materials are known to be capable of superconductivity—if the material is chilled to a temperature of about 450 degrees below zero Fahrenheit. That takes a fancy refrigerator. So superconductors have remained in the world of esoteric laboratory experiments since their discovery in 1911.

Until now, that is. In the last couple of years, materials have been found that don't have to be at 450 below to superconduct—they'll do it at 240 below. Still cold, but not nearly as cold as before.

Now the world's laboratories are competing to find some workable material that will superconduct at ordinary room temperature. Room-temperature superconductors could be used to make electric motors that would never overheat, power lines that would transmit 100 percent of the power from utility to customer, magnetic levitated trains, and superfast computers, among other things.

But today, experiment is ahead of theory in superconductivity. No one knows exactly how superconductors work. So no one knows how to discover better superconductors, except by trial and error.

"Superconductivity: The State That Came in from the Cold," *Science* 239:367 (January 22, 1988).

"The 1987 Nobel Prize for Physics," *Science* 238:481 (October 23, 1987).

"More Superconductivity Questions Than Answers," *Science* 237:249 (July 17, 1987).

DOES SMOKING CAUSE LUNG CANCER?

People who smoke tend to die of lung cancer more often than people who don't smoke. Does that mean that smoking causes lung cancer? Government reports on smoking and cancer mention five criteria for judging whether statistics prove a cause-and-effect relationship.

First: were the researchers who came up with the statistics biased, or could they have made some mistake? It takes several different studies, done by different people, at different times, in different places, using different methods, but reaching the same conclusions, to imply that the link between smoking and cancer is real.

Second: is the lung cancer death rate among smokers definitely greater than that among non-smokers, or is the difference so small that it could be a peculiarity of the group selected for the study?

Third: have the researchers figured in the possibility that something other than smoking—maybe heredity or smog—could be causing some of the cancer?

Fourth: does the smoking consistently happen before the lung cancer?

Fifth: does the idea that smoking causes lung cancer seem plausible in light of what's already known about smoking and about cancer?

As you can tell from the Surgeon General's warning on cigarette packs, the Public Health Service has concluded that smoking causes lung cancer. Eight long-term studies and dozens of smaller studies indicate that smokers are about ten times more likely to die of lung cancer than non-smokers—twenty or thirty times more likely in the case of heavy smokers. Lung tissue in people and animals starts to change after exposure to tobacco—and changes back after the tobacco is taken away.

U.S. Department of Health and Human Services, *Health Consequences of Smoking: Report of the Surgeon General—Cancer* (1982).
U.S. Department of Health, Education, and Welfare, *The Health Consequences of Smoking: A Reference Edition* (1976).

WHAT IS A GENE?

Consider the iris of the eye of a brown-eyed person. The color of the iris is one of the few inherited traits basically unaffected by diet and climate.

A brown iris is made of cells that contain brown pigment. The cells contain brown pigment because the genes in the nucleus of each cell told it to make brown pigment. A cell runs like an automated factory controlled by a computer program. If the computer program tells the factory to make brown pigment, the factory makes brown pigment. If the genes in a cell tell that cell to make brown pigment, the cell makes brown pigment.

Genes are made of a substance in the nucleus of a cell—nucleic acid, deoxyribonucleic acid, the famous DNA. Molecules of DNA take the form of fine strands. Every strand is made of smaller molecules connected end to end. There are millions of these smaller molecules in one strand, but they come in only four varieties. The four varieties of smaller molecules can be assembled in any order, like letters in a four-letter alphabet. A DNA strand is like a book written in a four-letter alphabet, with the text run together in one long line.

This book has different chapters. One chapter—or maybe several—carries the chemical formula for making brown pigment. Those chapters, those small sections of the DNA strand, are the genes for brown eye color.

"Genetics and Heredity," in *Encyclopaedia Britannica*, 15th ed.

"Eye Color," in Victor A. McKusick, *Mendelian Inheritance in Man: Catalogue of Autosomal Dominants, Autosomal Recessives, and X-Linked Phenotypes*, 4th ed. (Baltimore: Johns Hopkins University Press, 1975).

WHY DO CATS' EYES GLOW AT NIGHT?

You're driving along a lonely road at night. Ahead, in the dark, you see a pair of bright, disembodied eyes. You get closer and the eyes slip away into the grass—the eyes of a prowling cat.

The cat's eyes shine because your eyes, your car's headlights, and the cat's eyes are nearly in a straight line. The light from your headlights, or the sound from your engine, attracts the attention of the cat toward your car. The cat focuses its eyes on your headlights. In each of the cat's eyes, the lens brings the light from your headlights to a sharp focus, making a distinct image on the retina like the image on film in a camera.

But the light goes both ways. Some of it is reflected from the cat's retina back out through the lens of the eye. At night, the pupil of the cat's eye is wide open, so a lot of light goes through. The reflected light forms a narrow beam aimed at your car, because that's where the cat is looking. You look into the beam and get the weird impression that the eyes are shining by their own light. The impression is even more weird if you can't see the rest of the cat.

You can see the same glow in the eyes of rabbits that look up from their nighttime grazing to watch your car go by. And you can see it in the eyes of people in snapshots taken at evening parties. Anyone looking at the camera when the flash goes off has red eyes in the picture. The red comes from the color of the human retina. (This is easy to demonstrate with a model of an eye. Put a lens over a hole in a box whose depth is about the same as the focal length of the lens [the focal length should be marked on the metal barrel of the lens, e.g., 50mm]; stand at a distance and shine a flashlight beam on the box.)

But there's more. Cats have a special layer of tissue in their eyes that reflects light something like a metallic surface. It is located just behind the special cells in the retina that convert light to nerve impulses—the rod and cone cells. The apparent function of this reflecting layer is to make the cat's eye more efficient in dim light.

When light enters the cat's eye, some of it is absorbed by the rods and cones. But some light gets past the rods and cones and goes on to the back of the eye. The reflecting layer bounces this leftover light forward again, so it encounters the rods and cones a second time. Light entering a cat's eye has not one, but two chances to be detected by the cat.

Some light misses the rods and cones both times and leaves the cat's eye through the lens. That's the light we see as eyeshine. But because of this reflecting layer in the back of the eye, more light is used by the cat and less is wasted than if the reflecting layer weren't there.

Cats aren't the only animals that have reflective tissue in their eyes. Cattle, oxen, opossums, alligators, and some fishes are among the other animals that have it. We humans don't, and that's part of the reason we're not as good as cats at finding our way in the dark.

Marcel Minnaert, *The Nature of Light and Colour in the Open Air* (New York: Dover, 1954).

"Eye (Vertebrate)," in *McGraw-Hill Encyclopedia of Science and Technology* (New York: McGraw-Hill, 1987).

Stephen L. Polyak, *The Vertebrate Visual System* (Chicago: University of Chicago Press, 1957).

Paul Ehrlich, Dyes, and Drugs

A hundred years ago, clothes weren't available in nearly as many colors as they are now. But things were improving. Among the most exciting new chemical products of the nineteenth century were synthetic dyes, with names like mauve, amaranth, and Congo red.

A German medical student, Paul Ehrlich (1854–1915), was fascinated with what he learned about dyes in his anatomy classes in the 1870s. Just as some dyes stick to cotton but not to wool, some dyes stain only certain kinds of tissue, or certain parts of a cell, but not others. The dye methylene blue, for example, stains nerve cells but not other cells. So methylene blue highlights the nerve cells in a tissue specimen. There was what Ehrlich called a chemical affinity between the methylene blue and the nerve cells.

Now for Ehrlich's great leap of scientific imagination.

Maybe, he thought, a sick person or animal could be cured with a dye that would stick only to the bacteria causing the disease. If the right dye could be found and put into the bloodstream, it would attack the harmful bacteria like a magic bullet, leaving the regular cells untouched.

Ehrlich spent most of his career developing this idea. Early on, he found that the dye trypan red would cure South American horse disease. Later he developed the first safe and effective drug to treat syphilis in people.

Paul Ehrlich's work on the chemical affinity of dyes and his magic-bullet idea led all the way to sulfa drugs and other antibiotics still in use today.

Claude E. Dolman, "Paul Ehrlich," in *Dictionary of Scientific Biography* (New York: Charles Scribner's Sons, 1971).

Aaron J. Ihde, *The Development of Modern Chemistry* (New York: Harper and Row, 1964).

Alexander Findlay, *A Hundred Years of Chemistry* (Atlantic Highlands, N.J.: Humanities, 1965).

Of related interest: Allan M. Brandt, "The Syphilis Epidemic and Its Relation to AIDS," *Science* 239:375 (January 22, 1988).

Why you can never get to the end of the rainbow

It seems that anything as beautiful as a rainbow must lead to a wonderful place where it touches the ground. But if you've ever tried to approach a rainbow, you know that it recedes as you move toward it. The rainbow appears beyond that stand of trees, or over that next hill. When you move, the rainbow moves with you.

Sooner or later you realize that the end of the rainbow is not in any definite place you can mark on the map. No part of a rainbow is in any definite place, except in relation to your eye.

A rainbow is just a total of all the light coming to your eye from certain directions. The light from a rainbow is sunlight, reflected and broken into colors by water drops.

A rainbow always forms part of a circle, and the center of that circle is the point opposite the sun—from your point of view. The rule is that any water drops forty-two degrees of angle away from that point opposite the sun contribute to the rainbow you see.

Whether the water drops are ten feet away or ten miles away, they reflect light at that same angle and contribute to the same rainbow—for you. If you walk toward the end of the rainbow, it stays ahead of you as long as there are water drops in the air ahead of you.

Marcel Minnaert, *The Nature of Light and Colour in the Open Air* (New York: Dover, 1954).

H. Moysés Nussenzweig, "The Theory of the Rainbow," *Scientific American*, April 1977.

Robert Greenler, *Rainbows, Halos, and Glories* (New York: Cambridge University Press, 1980).

The difference between a square and a diamond

On red paper, trace the outline of a record album jacket, cut it out, and tape it to the wall with one side parallel to the floor. Everybody will say you have a square on your wall. The eye-brain combination is saying, in effect, "I see a shape with four equal sides and right angles at the corners. In school we called that a square. So I'm looking at a square."

Now take your red square and hang it by one corner. Bring in some new friends. They'll probably say you have a diamond on your wall. But a diamond also has four equal sides and right angles at the corners. Evidently the eye-brain combination looks at more than sides and angles; it says something like, "I see a four-sided shape with points on the top and bottom and sticking out of the sides. That's a diamond."

Now, with your diamond on the wall, tilt your head forty-five degrees and look. Your retina now has an image on it that's the same as the image of the red paper when you first hung it up. But it's still a diamond, not a square. Your earlier decision about what's top and what's bottom is still in effect, even though your head is still tilted.

So the task of judging a shape has at least two parts to it—the task of seeing sides and angles, and the task of assigning directions, such as top and bottom.

Sketch an outline of the continental United States, and turn the paper so the East Coast is down. If you show it to people without saying anything, how many will see a profile of Abraham Lincoln?

See Irvin Rock, "The Perception of Disoriented Figures," *Scientific American*, January 1974 (also bound in the *Scientific American* book *Recent Progress in Perception* [San Francisco: W. H. Freeman, 1976]).

Sounds over a Lake at Evening

Sometimes, standing by the lake shore or sitting in a rowboat just after sunset, you can hear voices from far away—real voices—with amazing clarity.

A good time to listen is just after sunset when the sky is clear and the wind calm. As night approaches, the water, no longer getting any sunlight, cools off. The cool water cools the air just above it. Meanwhile, a few dozen feet above the lake, the air not in contact with the water stays warmer.

A quarter of a mile away, someone speaks. The voice is carried by sound waves. You can think of a sound wave as an invisible wall of very slightly compressed air, traveling across the lake at the speed of sound.

But the speed of sound varies. Sound travels slightly faster in warm air than in cool air. The part of the invisible wall up in the warm air travels a little faster than the part down in the cool air near the water, so that the top gets ahead of the bottom. The invisible wall bends downward as it goes. Thus the sound of a distant voice is also bent downward. Less sound goes up into the sky at evening than at midday. Because of the cool surface air, evening sounds stay near the water and travel a long way—sounds of human voices, birds, bells, trains, motorboats, tape decks, and radios.

Joe R. Eagleman, *Meteorology: The Atmosphere in Action*, 2nd ed. (Belmont, Calif.: Wadsworth, 1985), chap. 13, "Atmospheric Optics and Acoustics."

Do THE BEST DOGS COME FROM THE POUND?

The best dog for someone else may not be the best dog for you. But there's something to the idea that a mutt from the animal shelter can give you the best of everything, including good health and good disposition. There's a principle of heredity at work.

A mutt is a dog whose parents are completely unrelated—they have no ancestor in common. Whatever undesirable genetic traits the parents may carry are likely to cancel each other out in the offspring.

On the other hand, if two parents are closely related—as brother and sister, for instance—if they have any recent ancestor in common, then each parent might have inherited a copy of the same bad gene from that common ancestor. If two dogs with the same bad gene now breed, some of the puppies will probably inherit two copies of that bad gene—and two copies are usually necessary for a genetic problem to show up. This can happen in puppy mills where incestuous matings are used to produce dogs in quantity for fast cash.

Responsible breeders breed only dogs whose ancestry they've checked out, looking for genetic problems. Responsible breeders also give their puppies individual care so the puppies learn to get along with people.

So some of the best dogs do come from the pound, and some of them come from top-notch breeders. Let the buyer beware of dogs from unscrupulous operators who use inbreeding for the sake of quantity, not quality.

The Monks of New Skete, *How to Be Your Dog's Best Friend* (Boston: Little, Brown, 1978).

Flip a Coin, Beat the Odds

In a hundred tosses of a coin, you expect about fifty heads. Ever tried it?
Imagine: a kitchen table, a chance idle moment. Flip a penny. Heads.
Toss again. Heads again? Somewhat unusual. Two heads is only one
of four possible results in two throws.

Heads again? There's only a one-in-eight chance of three heads in
three tosses.

Heads a fourth time! You've beaten one-in-sixteen odds! Interesting.
Keep flipping. Good luck.

Twenty heads in a row! No hidden magnets? The chance of twenty
straight heads is about one in a million. Success breeds success, right?

Wrong, in the case of *honest* coin flips. No matter what happened before, your chance for heads next time is always fifty-fifty.

Ninety-nine tosses, ninety-nine heads! The chance of that is about
one in six hundred billion billion billion. If you'd only known, you could
be rich!

At this point, some people might say you're on a roll. Others, that
you're due for a string of tails. Smart ones won't play. The chance of
getting heads next time is still fifty-fifty.

Two hundred and ninety-nine heads in a row? The odds against that
are about one with ninety zeros. Someone should see this—a TV reporter, a mathematician! Too late. The previous two hundred and
ninety-eight throws are in the past—gone.

Happy tossing. But remember: honest coin flips are independent. Past
performance has no bearing on future results.

The number of possible outcomes of n tosses of a coin is 2^n. Straight heads is only one of
the 2^n possibilities. 2^{20} = 1.049 x 10^6; 2^{99} = 6.338 x 10^{29}; 2^{299} = 1.109 x 10^{90}.

Stroboscopic Stagecoach Wheels

Maybe you've noticed something odd in western movies that show a
stagecoach leaving town. Even though the stagecoach is moving forward,
sometimes the wheels seem to turn backward. The spokes seem to be
going the wrong way.

To understand how this happens, remember that each second of movie film is really twenty-four pictures—or "frames," as they're called in the business—flashing onto the screen one after the other. Suppose that in the first frame there's a spoke on the stagecoach wheel that's in the twelve o'clock position, pointing straight up from the hub. By the next frame, the wheel will have turned a little. Now there may be a spoke in the eleven-fifty-nine position. It doesn't have to be the same spoke. In the third frame there might be a spoke photographed in the eleven-fifty-eight position; in the fourth frame, eleven-fifty-seven, and so on.

Since all the spokes look the same, when the pictures are run as a movie you see what looks like the same spoke moving from the twelve o'clock position, to eleven-fifty-nine, to eleven-fifty-eight, to eleven-fifty-seven, and so on. The wheel appears to turn backward.

There are variations on this as the stagecoach goes faster. The spokes appear to go backward, then forward, then backward, changing speed all the time: faster, then slower, then faster. The effect depends on where the spokes are when the camera catches them, and how each picture relates to the ones before and after. When the stagecoach gets going fast enough, each frame of the movie shows a blur instead of a clear image of spokes, and the illusion disappears.

If you don't like westerns, you can see the same effect in movies of propeller-driven airplanes starting their engines. The prop seems to go one way, then the other, always changing speed, until the effect disappears in a blur.

Marcel Minnaert, *The Nature of Light and Colour in the Open Air* (New York: Dover, 1954).

WHY ARE BELLS MADE OF METAL?

Why are bells made of metal? Or, to ask the question another way, why is metal a good material for bells?

Try a simple home experiment. Hold an ordinary stainless-steel kitchen table knife loosely between two fingers and tap it sharply with another knife. It rings like a bell.

When you tap on the knife, it flexes very slightly away from the place you tapped. Then it springs back. But here's the important part: the knife doesn't just spring back to its original shape. It springs back with so much energy that it overshoots and flexes the other way. And, having flexed the other way, the knife springs back yet again—with enough energy to overshoot the original position again, and flex the other way, again.

This flexing back and forth happens hundreds of times per second, and in a piece of the right kind of metal, it may go on for five or ten seconds or more. That's why metal is a good material for bells.

Each flex compresses the air near the knife. That's a sound wave. The rhythm of the flexing back and forth is regular, so the sound waves are regular and we hear a musical tone. In other words, the metal knife vibrates and makes a sound. A bell is a piece of metal in another shape doing the same thing.

Metal isn't the only kind of material that will ring like a bell. Glass also will do it. Some kinds of wood will ring for a short time—for example, the rosewood in xylophones and marimbas. Since these different materials share the property of ringing when struck, there must be something similar about the way the atoms in each material are held together.

Rodney Cotterrill, *The Cambridge Guide to the Material World* (New York: Cambridge University Press, 1985).

J. Bronowski, *The Ascent of Man*, book and video (Boston: Little, Brown, 1974).

Scientific American, September 1967.

Richard P. Feynman, *The Feynman Lectures on Physics* (Reading, Mass.: Addison-Wesley, 1964), vol. I, chap. 21.

WHY RIVERS DON'T FLOW IN A STRAIGHT LINE

If a river has a chance, it will meander, winding over the land in a series of loops. Geologists even call the loops meanders.

You can see meanders on a map—along the lower Mississippi, the Alabama River near Selma, the Arkansas near Tulsa, the Ohio near Evansville, and on thousands of smaller streams, wherever there's a steady flow of water over nearly flat land of fine-textured soils.

Rivers meander because any small bend in a river tends to grow.

Water flowing around a bend in a river is a little like a car speeding around a bend in a road. The water is thrown toward the outside of the turn. That fast-moving water erodes the riverbank on the outside of the bend.

Meanwhile, on the inside of the bend, the water flows more slowly. Sediment held by the water can settle out and accumulate along the inside bank.

So the water eats away the outside of the bend while it builds up the inside of the bend. The bend in the river grows into a big loop.

When the loop gets big enough, the water cuts across the narrowest part of it and starts the meandering process all over again. The cut-off loop becomes an oxbow lake.

A meandering river often alters its course. Mark Twain, in his book *Life on the Mississippi,* says that steamboat pilots of his day traveled up and down the river even when they weren't working, just to keep track of the latest changes in the course of the water.

See Luna Leopold and W. B. Langbein, "River Meanders," *Scientific American,* June 1966.

John S. Shelton, *Geology Illustrated* (San Francisco: W. H. Freeman, 1966).

What's Your Average Speed?

Strange noises are coming from your car. You are five hundred miles from home, at your favorite vacation spot. The drive out here took eight hours, for an average speed of sixty-two and a half miles an hour—fast, but probably illegal.

Anyway, you get a mechanic to look your car over before you start for home. The mechanic announces, "You have a worn clutch, bad piston rings, and a dirty carburetor, among other problems. Your car has only five hundred miles left on it!

"Furthermore, you'll be able to leave my garage at sixty-two and a half miles an hour, but your car will slow down steadily. Your speed will decrease to zero just as you crawl into your driveway. Your trip home will take you sixteen hours. Your speed will average out to thirty-one and a quarter miles per hour."

This mechanic remembers his math from school.

"During the first eight hours of your sixteen-hour trip, you'll cover three hundred and seventy-five miles," he says. "During the last eight hours, you'll go only one-fourth as far—a hundred and twenty-five miles. During the last four hours, you'll go about thirty-one miles. During your last half-hour you'll have plenty of time to wave hello to the neighbors, because you'll go only half a mile!"

"Hmm—maybe I should fly home. There's a commuter flight that takes two hours."

"Better count on three hours at the airport here, and another three hours to get your luggage at the other end," the mechanic says.

"Gee, that's a total of eight hours—if I fly home, my average speed will be the same as if I had a good car! Maybe I can sell this one for scrap and get enough cash to buy a plane ticket."

THE BASIC UNIT OF LIFE

One of the most important unifying ideas in modern biology is that the cell is the basic unit of life. All living things are composed of one or more cells.

The basic role of cells has gradually become apparent since the invention of the microscope nearly four hundred years ago. Cells were seen first in plants; then the basic similarity between plant and animal cells was noticed; still later it became clear that new cells come only from other cells.

Biologists have been impressed by the similarity of cells, whether those cells come from oak trees or human beings. All cells are surrounded by a membrane that controls what gets into and out of the cell. All cells are chemical factories, taking in nutrients, making new substances, changing energy from one form into another, and eliminating wastes. All cells contain genetic information, coded in the form of molecules of deoxyribonucleic acid, the famous DNA.

Cells also contain strange relics of the distant past. Within the cells of algae, plants, and animals, there are little bodies that have their own DNA, their own genetic system, their own line of inheritance separate from the rest of the cell. These little objects go by the names of mitochondria and chloroplasts. Algae, plants, and animals, including us, have mitochondria in their cells; only algae and plants have chloroplasts. Mitochondria and chloroplasts are indispensable—they supply chemical energy to the rest of the cell. But in many ways they resemble bacteria living in the cell as guests. Evidence now indicates that those mitochondria and chloroplasts are, in fact, descendants of bacteria that lived independently a billion years ago.

C. P. Swanson and P. L. Webster, *The Cell*, 5th ed. (Englewood Cliffs, N.J.: Prentice-Hall, 1985).

Helena Curtis, *Biology*, 4th ed. (New York: Worth, 1983).

How Aspirin Got Its Name

The word "aspirin" was invented in Germany just before the turn of the century, at the chemical firm founded by Friedrich Bayer.

Felix Hoffmann, a staff chemist at Bayer, was studying the usefulness of salicylic acid, a substance that had been on the market for years as a food preservative. Hoffmann wondered whether it could be used as a drug. Chemicals related to salicylic acid were present in willow bark and oil of wintergreen, which had been used as traditional pain relievers.

Hoffmann tried giving pure salicylic acid to sick people. It relieved pain and reduced fever, but it was hard on the stomach. Hoffmann modified the pure acid by a process called acetylation, hoping that acetylated salicylic acid would be less upsetting but still effective. It was. Hoffmann had created a potential best-seller among drugs. But would the public swallow the name—acetylsalicylic acid?

One of Hoffmann's superiors at Bayer, Heinrich Dreser, had an idea. Dreser remembered that salicylic acid was also found in plants of the type known as spirea; in that form, the stuff was called spiric acid. Dreser combined "a" for "acetylated" with "spirin" for spiric acid to create the now-famous word "aspirin." That was in 1899, and that's how aspirin got its name.

Who knows what hard-to-pronounce scientific terms of today will be streamlined to become household words of tomorrow?

Carrie Dolan, "What Soothes Aches, Makes Flowers Last, and Grows Hairs? Aspirin, the Model-T of Drugs, Does All This and More; Fire Ants and Willow Bark," *Wall Street Journal*, February 19, 1988.

Carroll Hochwalt, "The Story of Aspirin," *Chemistry* 30(6):10 (1957).
Aaron J. Ihde, *The Development of Modern Chemistry* (New York: Harper and Row, 1964).

COLD WATER AT THE BOTTOM OF A LAKE

It's a warm summer morning, and you're standing in a freshwater lake with water up to your neck. You notice that your feet are colder than your shoulders. Cold water sinks, of course. But how much colder could the water be on the very bottom of the lake?

It depends on the lake—where it is, and how deep it is. But in a freshwater lake the coldest water at the bottom will not be colder than 39 degrees Fahrenheit.

Fresh water is denser at a temperature of 39 degrees than at any other temperature. Any 39-degree water in a lake will go to the bottom. Water that's warmer or colder than 39 degrees will float on top and be exposed to the weather.

So the water in a lake usually divides into layers according to temperature—except for two brief periods each year when the surface water temperature matches the deep water temperature. The temperatures match for a few days in spring as cold surface water is warmed up by the sun, and in fall when warm surface water cools off with the approach of winter. During those special times, called the spring turnover and the fall turnover, all the water in the lake has the same temperature.

Then, wind stirs the water, mixing oxygen and nutrients through the whole lake. For fish and other things living in a lake, those turnovers are among the year's biggest events.

Robert Leo Smith, *Ecology and Field Biology* (New York: Harper and Row, 1974).
George K. Reid, *Pond Life* (New York: Golden Press, 1987).

WHY DO WOMEN LIVE LONGER THAN MEN?

In the industrialized world, women live longer than men, on the average. And women live longer even though the game of life, if you want to call it that, starts with men in the lead.

It's estimated that at conception males outnumber females by about

115 to 100. In the months before birth, the male lead is reduced by miscarriages and stillbirths. At birth, males outnumber females by about 105 to 100.

By age thirty, males and females are about equal in number. After thirty, women pull ahead in the survival game. By age sixty-five, women outnumber men by about 120 to 100.

The complex reasons for this are just beginning to be figured out.

Men are more likely to die of heart disease, stroke, lung cancer, accidents, and homicide. Women tend to be sicker, but their diseases are less likely to be fatal—diseases like arthritis, lupus, and sinusitis.

Women seem to be protected from heart disease by their sex hormones, known as estrogens. Estrogens somehow keep the level of harmful blood cholesterol down. On the other hand, men's sex hormones, the androgens, tend to raise their harmful cholesterol levels. This jibes with the observation that while men have a big increase in their chance of heart disease during their forties, the increase doesn't hit women till after menopause.

All this is changing, though—in particular, more women are smoking now than decades ago. Maybe women won't outlive men so often in the future.

"Why Do Women Live Longer Than Men?" *Science* 238:158–160 (October 9, 1987).

A THIRTY PERCENT CHANCE OF RAIN

If a weather forecast includes a thirty percent chance of rain, what are we supposed to do—carry thirty percent of an umbrella?

A thirty percent chance of rain means that the weather forecasters have combined all their knowledge of the history and present state of the atmosphere and have concluded that out of a hundred days like this one, about thirty will have rain.

You'll notice that whether or not it actually rains today, the forecasters may still be right. Have they evaded responsibility by refusing to give a yes-or-no answer?

Not exactly. The forecasters have used the thirty percent figure to describe the imperfection of their knowledge as accurately as they can.

Whether we take an umbrella depends on how much we care about getting wet. We ask ourselves questions that only we can answer:

"How many hours do I expect to spend outdoors today?"

"If I leave my umbrella home and it rains, will I be extremely annoyed, only slightly inconvenienced, or just amused?"

"If I take my umbrella and it doesn't rain, will I leave the umbrella somewhere by mistake and lose it?"

"Am I carrying anything today that must not be allowed to get wet, like a watercolor painting or a cat?"

The forecasters have done their part; they've given their best estimate of the chance of rain. We have to decide for ourselves, based on our own values, whether or not to take a hundred percent of an umbrella. If rain does fall, thirty percent of an umbrella won't keep us dry.

For more information on scientific risk assessment, see Richard Wilson and E. A. C. Crouch, "Risk Assessment and Comparisons: An Introduction," *Science* 236:267–270 (April 17, 1987).

Limeys

The derogatory slang epithet "limey" is short for "lime-juicer."

The original lime-juicers were British sailors of the 1800s who got lemon or lime juice with their food in order to prevent scurvy, a condition characterized by rotten gums, weak knees, and fatigue. During the late 1700s, about one-seventh of the sailors of the British navy were disabled by this disease.

A Scottish naval surgeon, James Lind, collected information about scurvy and learned that it had often been cured by a diet of fresh fruits and vegetables—which the gruel-eating sailors certainly weren't getting. Lind understood that it was impractical to carry a lot of fresh produce on a ship in those days. But he experimented and found that the juice alone from lemons and limes and oranges could cure and prevent scurvy. Thus he recommended that sailors drink lemon juice at sea. The navy eventually took Lind's advice and put lemon juice aboard British ships starting in the 1790s. By the mid-1800s limes were cheaper than lemons, so lime juice was used instead. The British sailors became "lime-juicers," then "limeys."

Today we'd say that what the British sailors were getting from the fruit juice was vitamin C. For a while vitamin C was called the antiscorbutic substance because it prevented scurvy. The streamlined generic name ascorbic acid was invented in 1933.

Eric Partridge, *A Dictionary of Slang and Unconventional English*, ed. Paul Beale (London: Routledge and Kegan Paul, 1984).

"James Lind," in *Dictionary of Scientific Biography* (New York: Charles Scribner's Sons, 1973).

"Biochemical Components of Organisms" and "Medicine," in *Encyclopaedia Britannica*, 15th ed. (1986).

How DOES LUNG CANCER START?

Exactly how does lung cancer start? There's evidence that it happens something like this:

Your lungs are made of cells. When you're growing up, your lung cells often divide to make new lung cells. Normally this process is magnificently controlled and coordinated to produce a growing, healthy lung.

Once you've grown up, your lung cells don't divide nearly so often. They don't divide because there are genes in each cell that keep the cell from dividing—genes whose purpose is to say to the cell, "Don't divide."

Now, suppose some of the genes in a lung cell are damaged, maybe by something you inhale—something in tobacco smoke, for instance. Suppose the "don't divide" genes are the ones that get damaged. Then the cell will go ahead and divide. There's nothing stopping it. It will divide into two new lung cells, each of which has damaged "don't divide" genes, just like the original. Each of those will, in turn, divide into two cells—with damaged "don't divide" genes—and so on, to uncontrolled copying of the original bad cell. That's lung cancer.

This connection between a gene not doing its job and cancer has been seen with particular clarity in the case of lung cancer, but it may be the central process in other kinds of cancer as well.

J. Michael Bishop, "Oncogenes," *Scientific American*, March 1982.

"Single Gene Deficiency Linked to Lung Cancer," *New York Times*, December 11, 1987, p. 17 (national ed.).

K. Kok et al., "Deletion of a DNA Sequence at the Chromosomal Region 3p21 in All Major Types of Lung Cancer," *Nature* 330:578 (December 10, 1987).

H. Brauch et al., "Molecular Analysis of the Short Arm of Chromosome 3 in Small-Cell and Non-Small-Cell Carcinoma of the Lung," *New England Journal of Medicine* 317:1109–1113 (October 29, 1987).

THE LIGHT OF SPRING

When people talk about springtime, they usually talk about warm weather or melting snow. But for most living things, the real sign of spring is increasing sunlight.

The sun is higher in the sky at noontime on a spring day than at noon on a winter day. And the sun is up for more hours each day as winter turns into spring and then summer.

Green plants use sunlight to make their food. Each plant has a strategy for getting the light it needs.

In a deciduous forest, the major consumers of sunlight are the big trees—oaks, maples, hickories, and so on. By summertime, these trees make a canopy of leaves that intercepts most of the sunlight reaching the forest. How do other plants survive?

Beneath that canopy, plants have to get by on a tiny fraction of the intensity of full sunlight. Some of these shade plants are the Jack-in-the-pulpit, wild ginger, and trilliums.

Other plants survive on the forest floor by taking advantage of the full sunlight of springtime, when the sun is high in the sky but not yet blocked out by the leaves of the big trees. The so-called ephemeral wildflowers, including the trout lilies, Dutchman's breeches, and spring-beauty, bloom and die back before the tree leaves are fully expanded. These ephemeral wildflowers have to make enough food during their short growing season to last the rest of the year, storing it in a bulb or some other underground organ.

So ephemeral wildflowers, shade plants, and big trees have a schedule for sharing sunlight. That pattern of coexistence must have taken a long time to develop.

John Mitchell and the Massachusetts Audubon Society, *The Curious Naturalist* (Englewood Cliffs, N.J.: Prentice-Hall, 1980).

Peter Farb, ed., *The Forest* (New York: Time, 1961).

Robert Leo Smith, *Ecology and Field Biology* (New York: Harper and Row, 1974).

VITAMINES AND VITAMINS

In 1912 a Polish biochemist, Casimir Funk, published an article about food substances that could prevent diseases like beriberi and scurvy. Funk's analysis showed that these disease-preventing food substances might be members of a family of chemicals called amines. These substances were vital for a healthy diet, so Funk called them vital amines, or vitamines—spelled like "vitamins," but with an "e" at the end.

By 1916 there was evidence that these disease-preventing food substances might not actually be what chemists call amines. The name "vitamine" was thrown out in favor of the names "fat-soluble A" and "water-soluble B." (Only those two types were known at the time.)

In 1920 another chemist wrote that the names "fat-soluble A" and "water-soluble B" were unwieldy. He suggested dropping the "e" from the old word "vitamine" and calling the substances vitamins. Whatever the substances might turn out to be, the name "vitamin" would be chemically permissible. As new types were discovered, they could be given letters—vitamin A, vitamin B, vitamin C, and so on.

See "Vitamin," in *Supplement to the Oxford English Dictionary*, and "-in" in the *OED*.
Aaron J. Ihde, *The Development of Modern Chemistry* (New York: Harper and Row, 1964).
Elmer V. McCollum, *A History of Nutrition: The Sequence of Ideas in Nutrition Investigations* (Boston: Houghton-Mifflin, 1957).

COLORS AND THEIR OPPOSITES

Let's begin with some thoughts about paint.

We usually think of paint as a substance that adds color to things. But from a physical point of view, paint works by taking colors away—from white light.

Take the example of a red car parked in the sun. The sunlight already contains all the colors of the rainbow. The red paint on the car absorbs the non-red colors from sunlight and reflects only red to our eyes. Red paint absorbs, or subtracts, non-red from white.

What happens to those non-red colors absorbed by the paint? Actually, the absorbed light is converted to heat. But here's an almost philosophical question. If somehow we could see the light that red paint absorbs, what would we see? That is, what does non-red look like?

In fact, it's a bluish-green color called cyan. People who work with color professionally say that cyan is the complement of red.

Cyan is the color that red paint takes away from white light. Now, what if you put cyan and red back together? When you add cyan light to red light, you get white light—you're reassembling the colors of sunlight. Whatever colors are not supplied by the cyan are supplied by the red.

With paint it's different. Cyan paint added to red paint makes black paint, because whatever colors the cyan doesn't absorb, the red does. Nothing is reflected, so the paint looks black.

The mixture of any color and its complement makes white if you're mixing *light,* and black if you're mixing *paint.* This is one of the basic secrets in any business involving color—from painting to color television.

See "Colour" and related entries in *Encyclopaedia Britannica.*
Matthew Luckiesh, *Color and Its Applications* (New York: D. Van Nostrand, 1915).

How DOES THE WORLD LOOK TO A BEE?

To describe light in a general way, you need to specify at least three qualities: its brightness or intensity, its color, and its polarization.

Polarization is a quality our eyes don't detect. We have no everyday words to describe polarization, so we have to resort to a more or less scientific description of it.

If we think of light as a wave traveling through space something like a ripple crossing a pond, then we can think of polarization as describing the direction in which the wave vibrates. The vibration in a light wave is always perpendicular to the direction the wave is traveling. But the vibration of light can be up and down, sideways, or any combination of the two.

If the vibrations are in random directions, the light is said to be unpolarized; if all the vibrations are in the same direction, it's completely polarized; in-between amounts of polarization are most common.

To our eyes, polarization makes no difference. But it has been known for decades now that insects in general, and bees in particular, can detect the direction a light wave is vibrating. Bees navigate by referring to the direction of the sun. But they don't have to see the sun directly; all they need is a clear view of a small piece of the sky. The blue glow of the sky is polarized, and the direction and the amount of polarization are different in every part of the sky, depending on where the sun is. A bee can tell where the sun is by looking at the polarization of any small piece of the sky.

So bees have a dimension to their vision that we lack. In addition to color and brightness, bees see polarization. What does that sensation feel like? How does the world look to a bee? We can only wonder.

G. P. Können, *Polarized Light in Nature* (New York: Cambridge University Press, 1985).

Knut Schmidt-Nielsen, *Animal Physiology: Adaptation and Environment*, 3rd ed. (New York: Cambridge University Press, 1983).

Marcel Minnaert, *The Nature of Light and Colour in the Open Air* (New York: Dover, 1954).

DEATH OF THE DINOSAURS: A QUICK REVIEW

The fossil record indicates that about 65 million years ago, more than half of all living species, including the last of the dinosaurs, suddenly became extinct. What happened? This mystery is one of the best-known in all science.

The modern controversy began in 1980 with the discovery of a layer of clay, enriched in the element iridium, at a depth in the rocks corresponding to the time of the extinction. The layer was found at several places around the world. Iridium occurs in greater concentrations in meteorites than in earth rocks, so Luis Alvarez at the Lawrence Berkeley Laboratory in California suggested that Earth might have been struck by a giant meteorite or a small comet 65 million years ago. The impact would have raised a worldwide dust cloud, blocking the sun, leading to the death of green plants and of animals who ate them. More recent evidence in favor of the impact idea includes the discovery of quartz grains from the period showing signs of shock.

But some geologists argue that the iridium enrichment and the shocked quartz could be explained by volcanic eruptions instead of an impact. And some paleontologists say that the extinctions 65 million years ago may not have happened at exactly the same time. Also, there were many survivors of the extinction event, including the ancestors of all organisms now living.

Maybe, a more recent argument goes, there were many smaller impacts over a period of several hundred thousand years, spreading the extinctions out over time. Maybe Earth was hit by a swarm of comets thrown into the inner solar system by an encounter with a nearby star!

These are just a few highlights of the dinosaur-extinction controversy. No resolution is in sight.

"Star-Struck? Impacts' Role in the History of Life Remains Contentious," *Scientific American*, April 1988, p. 37.

WHY DO WE PUT CUT FLOWERS IN WATER?

Water keeps cut flowers and other plants crisp because of one of the most important and all-pervasive natural processes operating on the face of planet Earth: osmosis. Osmosis is the process in which liquid water tends to move toward regions with a higher concentration of dissolved substances. The dissolved substances might be minerals, sugars, anything— water will usually move to where there's more dissolved material.

Each cell of a plant has a sort of skin—a membrane. Water can pass through the membrane easily, but other materials can't. Each plant cell maintains a relatively high internal concentration of dissolved materials. Water therefore tends to move into the cell.

As long as the concentration of dissolved substances is higher inside the cell than outside, water will usually push its way in. The water pressure that builds up inside the cell is what gives a healthy plant its crisp texture. Often that pressure will make cells expand; that's one way plants grow. And from this you can see why plants wilt if they don't get enough water: the cells lose internal water pressure.

So plants in general, and cut flowers in particular, stay crisp because dissolved materials in effect draw water through the cell membranes into the plant cells by the process of osmosis. There are other ways living things move water from one place to another, but osmosis is one of the most important.

Frank B. Salisbury and Cleon W. Ross, *Plant Physiology*, 3rd ed. (Belmont, Calif.: Wadsworth, 1985).

Helena Curtis, *Biology*, 4th ed. (New York: Worth, 1983).

LIFE WITHOUT ZERO

Zero is one of humanity's greatest inventions, a symbol that stands for nothing in a very definite way.

The ancient Greeks and Egyptians had no zero. They used completely different symbols for 9, 90, 900, and so on. This system has a couple of big disadvantages. First, it has symbols only for numbers people have already thought of. If you want to talk about, say, 900 billion, you will have to invent a symbol for it. The old Greek and Egyptian systems also make arithmetic hard. Without zero, multiplying 3 times 90 is a whole different problem from multiplying 3 times 9.

The first known zero symbols appear in Babylonian clay tablets of about 500 B.C.; there, the zero was used to clarify the symbols for large numbers.

The idea that zero can be treated in arithmetic problems as a number, like any other number, came from a Hindu astronomer of the seventh century A.D., Brahmagupta. He was the first to write down the rules for arithmetic with zeros. Western civilization didn't adopt arithmetic with zeros until about seven hundred years later, based on the work of the thirteenth-century Italian mathematician Leonardo Fibonacci.

Thanks to zero, we have to learn multiplication tables only up to ten times ten. Thanks to zero, we can punch any number into our calculators using just ten keys. And if we want to imagine some gigantic number, we can do it easily—just by adding more zeros.

"Zero," in *McGraw-Hill Encyclopedia of Science and Technology* (New York: McGraw-Hill, 1987).

"Numerals and Numeral Systems," in *Encyclopaedia Britannica*.

O. Neugebauer, *The Exact Sciences in Antiquity* (New York: Dover, 1968).

"Brahmagupta," in *Dictionary of Scientific Biography* (New York: Charles Scribner's Sons, 1970).

Prostaglandins

Prostaglandins are a family of chemical messengers in the body. In that respect they're like hormones. But unlike the familiar hormones, prostaglandins don't come from special glands—apparently they can be made by cell membranes in just about any part of the body.

Prostaglandins are found, among other places, in human semen. A remarkable early discovery was that prostaglandins in semen induce muscle contractions in the uterus, helping sperm to be carried into the female reproductive system. So in this case, prostaglandins from one individual seem to have their function in another individual.

Other prostaglandins have other functions. Some influence the secretion of digestive juices. Some cause blood vessels to constrict, others cause them to dilate; some help to promote blood clotting, others help to prevent it.

A byproduct of the study of prostaglandins was an answer to the decades-old mystery of how aspirin works. Various prostaglandins in large amounts can cause tissue inflammation, headache, and fever, as well as muscle contractions. About 1970 it was found that aspirin inactivates an enzyme necessary for the production of prostaglandins. Apparently that's why aspirin relieves inflammation, headaches, fever—and menstrual cramps. (That research led to a Nobel Prize in 1982.)

Prostaglandins were discovered in 1930, but not much could be learned about them because they're made by the body in only tiny amounts and are broken down by enzymes in a matter of minutes. Progress in our knowledge about them came more rapidly with new techniques for sensitive chemical analysis developed in the 1960s and beyond. But much about prostaglandins is still unknown.

Helena Curtis, *Biology*, 4th ed. (New York: Worth, 1985).
J. R. Vane, "Inhibition of Prostaglandin Synthesis as a Mechanism for the Action of Aspirin-Like Drugs," *Nature: New Biology* 231:232–235 (1971).
J. B. Smith and A. L. Willis, "Aspirin Selectively Inhibits Prostaglandin Production in Human Platelets," *Nature: New Biology* 231:235–237 (1971).

Sweetened Condensed Milk

Sweetened condensed milk is a good ingredient for sweet recipes because of all the added sugar—about 25 percent by weight. But when sweetened condensed milk was invented in the 1800s, the original reason for adding sugar to the milk was not for flavor, but for protection against spoilage. And it works—even after you open the can, sweetened condensed milk keeps longer than fresh milk.

That added sugar kills bacteria that otherwise would digest the milk and spoil it. The sugar kills not by poisoning the bacteria, but by a more direct physical process. It draws water out of the bacteria so the bacterial cells shrivel and die.

Each bacterial cell has a sort of skin—technically, a membrane. Water can pass through this membrane pretty easily, but substances dissolved in the water can't. Water has a natural tendency to move toward any region where there's a high concentration of dissolved substances. A bacterial cell in a can of sweetened condensed milk finds itself immersed in an extremely concentrated solution of sugar. Water inside the cell will therefore pass out through the cell membrane into the sugar solution. The bacterial cell dehydrates and dies in a sea of sugary water.

Sugar added to fruit has the same effect—that's the idea behind fruit preserves. Other foods are preserved with salt, exploiting the same principle.

That tendency of water to move toward a region where there's a high concentration of dissolved substances goes by the technical name of osmosis. In living things osmosis is one of the most important ways water gets from one place to another.

Harold McGee, *On Food and Cooking: The Science and Lore of the Kitchen* (New York: Macmillan, 1974).

Irma S. Rombauer and Marion Rombauer Becker, *The Joy of Cooking* (Indianapolis: Bobbs Merrill, 1974).

WHAT THE WEATHER REPORT DOESN'T TELL YOU

When we hear a report that the temperature outside is, say, 70 degrees, we're usually hearing a measurement made with a thermometer at least six feet off the ground. But there can be big variations in temperature closer to the ground. On a sunny day, it may be five or ten degrees warmer at ankle level than at eye level. A cool day for us can be a warm day for rabbits and squirrels.

Sunshine warms both the ground and the air, but it warms the ground better than it warms the air. Soil and green plants absorb sunlight and convert it to heat better than air does. So a field of grass and clover and dandelions feels warm to the touch on a cool, sunny afternoon.

The warm ground heats the air a few inches above the ground. That surface air doesn't move much—it's not affected by breezes—so it gets warm and stays warm.

Where there's vegetation, the level of warmest temperature tends to follow the height of the leaves. The warmest place in a field of clover may be about half an inch above the ground; in a field of mature corn, four or five feet up; in a forest of oaks and maples, the hottest place is at the top of the canopy of leaves—maybe a hundred feet up.

This principle applies even if the vegetation is plastic. At a daytime major-league baseball game, the temperature might be 90 in the stands and 110 on the field, partly because artificial grass is so good at changing sunlight to heat.

Robert Leo Smith, *Ecology and Field Biology* (New York: Harper and Row, 1974).

Food and Agriculture Organization of the United Nations, *Forest Influences: An Introduction to Ecological Forestry*, Forestry and Forest Products Studies no. 15 (New York: Unipub, 1962).

WHY MOWING THE LAWN DOESN'T KILL THE GRASS

If you cut down an oak tree, the stump dies. But if you cut grass, you don't hurt it at all. That's because new growth on an oak tree is at the tips of the branches; new growth in grass happens at ground level. Also, grasses, unlike other plants, can replenish their leaves.

A blade of grass is the end of a long, narrow leaf. If you trace a grass blade back to the stem on a tall grass plant, you see that the blade comes from a sheath wrapped around the stem. At the base of that sheath is a node, a place where the stems of some grasses have a slight bulge. Nodes are where new growth happens in a grass plant. A short grass plant has at least one node near ground level, out of reach of the lawn mower; a tall grass plant may have several more nodes farther up along the stem.

When a grass blade is cut off by a lawn mower or a grazing animal, some as-yet-unknown signal is sent down to the node, stimulating it to produce more leaf. Grazing animals take particular advantage of this; eating the grass causes more to grow in its place.

The capacity to add new material to old leaves is characteristic of grasses. Other plants generally don't have this ability. An oak tree grows new leaves every year, but it can't replace part of an existing leaf. An oak leaf grows to a certain mature size and stops. If part of an oak leaf is cut away, it doesn't grow back.

The ultimate fate of an oak leaf is old age and death; leaves of grass remain youthful all summer long.

Frank B. Salisbury and Cleon W. Ross, *Plant Physiology,* 3rd ed. (Belmont, Calif.: Wadsworth, 1985).

Helena Curtis, *Biology,* 4th ed. (New York: Worth, 1985).

A MIRROR RIDDLE

Exactly what is the difference between the appearance of a real object and its reflection in a mirror?

Obviously, a mirror reverses the image of an object in some way. For instance, when you look into a bathroom mirror, you see an image with left and right switched. If you hold up a toothpaste tube, the letters on the reflected tube are backward. Evidently, reflection in a bathroom mirror reverses left and right.

But not all reflections reverse left and right. Some switch top and bottom. Think of trees on the far side of a lake, and how they're reflected in the water. The reflected trees have top and bottom switched, but left and right remain the same for the reflected trees and the real trees.

It seems like a contradiction: a bathroom mirror switches left and right, and the surface of a lake switches top and bottom. How can that be? (The explanation, by the way, has nothing to do with the difference between glass and water. If you put a mirror flat on a table, the reflections will be oriented the same way as in the surface of a lake.) Is there some precise way of describing the essential difference between a real object and its mirror image—some rule that will work in every situation? Here's a hint: look at the reflection of a clock in a mirror—the old-fashioned kind of clock, with hour and minute hands. A second hand is even better. What's the essential difference between the real clock and its reflection?

Think about what you saw when you looked at the reflection of the clock. While the hands on a real clock run forward, or clockwise, the hands on the reflected clock run backward, or counterclockwise. Therein lies the answer to our riddle: a mirror switches the clockwise and counterclockwise directions. Whether the reflection is right side up, upside down, or sideways simply depends on where you put the mirror. But the essential difference between reality and reflection is the reversal of clockwise and counterclockwise.

Look at the letter "p," as in "toothpaste" on your toothpaste tube. The shortest trip from the top of that letter "p" around the loop is clockwise on the real tube and counterclockwise in the upside-down reflection in the water. On the face of your clock, the shortest trip from the 12 to the 3 is clockwise in reality and counterclockwise in the mirror.

A mirror switches clockwise and counterclockwise. That surprisingly subtle rule is the only one that always works, regardless of the position of the mirror, or the person looking, for any reflected image.

GALILEO'S JOB APPLICATION

Here's a story about patronage through flattery: how a seventeenth-century scientist made his living.

Galileo Galilei wanted a court appointment with those well-known patrons of the arts, the Medici family of Florence. In 1607 word came that Prince Cosimo de Medici was interested in magnetism. Galileo responded with a flattering and educational gift: a lodestone, a natural magnet, on a handsome base inscribed with the Latin words *vim facit amor*, "love produces strength." The lodestone's strength was demonstrated by two little iron anchors stuck to its poles—anchors made under Galileo's personal supervision by artisans in Venice.

Galileo's pressing concern now was to deliver this lodestone to the Medici courier, whose address he had been given, in time for the last Sunday night dispatch from Venice to Florence. His own letters give us a picture of Galileo quite different from those serene museum portraits: the great astronomer, plying the canals of Venice on a rainy Sunday night with a surly gondolier, knocking on one door after another in the dark, looking for the Medici courier.

The lodestone eventually made it to Florence, and Galileo got his appointment, not so much because of the lodestone but because he named the four large moons of Jupiter (which he discovered) the Medicean stars. The name didn't stick, however. Today we call Jupiter's four largest moons the Galilean satellites.

Richard S. Westfall, "Science and Patronage: Galileo and the Telescope," *ISIS* 76:11–30 (1985).

THE CONSEQUENCES OF SMALLNESS

Put a teaspoonful of granulated sugar in a glass of water and stir. The sugar dissolves in a few seconds. Do the same thing with a single lump of very hard candy, and two or three minutes later some of the candy will still be undissolved. The amount of sugar is about the same in both cases. The difference is that the small size of the grains of granulated sugar give the sugar and water more opportunity to interact.

Dissolving happens only where sugar meets water. The more square inches of sugar exposed to water, the faster the sugar dissolves. In general, thousands of small grains have more surface area than one big lump of the same volume. It's amazing how much surface area you can get by grinding a solid lump into a fine powder. One cubic inch of material, divided into particles a hundred-thousandth of an inch wide, has a surface area of several hundred square yards.

This has many implications—in biology, for instance. Large living things are made of many tiny cells. Each cell must constantly adjust its internal balance of water, nutrients, and waste products to function properly. These substances get into and out of cells through their surfaces. The more square inches of cell surface a living organism has, the faster it can adjust its chemical balance. So it's advantageous to be made of small cells.

William A. Kieffer, *Chemistry: A Cultural Approach* (New York: Harper and Row, 1971). Helena Curtis, *Biology*, 4th ed. (New York: Worth, 1983), p. 98.

OZONE AND ICE

Ozone is a natural component of the Earth's upper atmosphere. A molecule of ozone is made of three oxygen atoms bound together; a molecule of the oxygen we breathe has two oxygen atoms. That's an important difference. The three-oxygen molecule blocks the sun's ultraviolet light, which would disrupt life on Earth if it reached the ground in large amounts.

Ozone in the upper atmosphere has been disappearing at an alarming rate in the last twenty years, especially over Antarctica. The culprit seems to be chlorine from chlorofluorocarbons, man-made chemicals once used in spray cans and still used in air conditioners and foam cushions. Chlorine destroys ozone by pulling loose one of its three oxygen atoms.

There's a natural process that was once thought adequate to protect ozone from attack by chlorine. In this process, the harmful chlorine is trapped by chemical reactions with naturally occurring compounds containing nitrogen. But there's yet another factor in the process: ice crystals in high-altitude clouds that form during the long, cold Antarctic winter night. Apparently these crystals trap the helpful nitrogen compounds.

So, chlorine from chlorofluorocarbons destroys ozone. Natural nitrogen compounds can prevent that destruction, but not if those nitrogen compounds are trapped in ice clouds. And that may be only part of the story. A cloud of tiny ice crystals has an immense amount of surface area where unknown chemical reactions might happen. Chemistry on the surface of a solid particle is different from chemistry in a test tube, and less is known about it. More needs to be learned about the chemistry of ice clouds.

Richard A. Kerr, "Stratospheric Ozone Is Decreasing," *Science* 239:1489–1491 (March 25, 1988).
Richard Monastersky, "Decline of the CFC Empire," *Science News* 133 (April 9, 1988).
Richard A. Stolarski, "The Antarctic Ozone Hole," *Scientific American*, January 1988.

Benjamin Franklin's Madeira Wine Surprise

Benjamin Franklin, who discovered that lightning is electricity as well as helped to write the Constitution, also speculated about the boundary between life and death. It was a great mystery in his day, as it is in ours.

One day when Franklin went to dinner at the home of some friends in London, he took along a bottle of Madeira wine from Virginia. At the table the bottle was opened and, as Franklin tells it, "three drowned flies fell into the first glass that had been poured."

Franklin doesn't say how his friends reacted—maybe finding flies in your drink was common in those days. Franklin, however, saw this as an opportunity to test an idea he'd heard—the idea that drowned flies could be revived by sunlight.

He strained the flies out with a sieve and put them in the sun. After three hours, according to Franklin, two of the three flies revived and flew away, "finding themselves in Old England," he wrote, "without knowing how they came thither."

"I wish it were possible," he went on, "from this instance, to invent a method of embalming drowned persons, in such a manner that they may be recalled to life at any period, however distant; for having a very ardent desire to see the state of America a hundred years hence, I should prefer to any ordinary death the being immersed in a cask of Madeira wine, with a few friends, till that time, to be then recalled to life by the solar warmth of my dear country!"

Benjamin Franklin, writing in 1773. We don't know exactly what happened to those flies, but we can marvel at Franklin's anticipation of speculative fiction of the twentieth century.

From a letter from Franklin to Barbeu Dubourg, published in N. G. Goodman, ed., *The Ingenious Dr. Franklin: Selected Scientific Letters of Benjamin Franklin* (1931; Philadelphia: University of Pennsylvania Press, 1956), pp. 150–152.

Viruses Attack Bacteria

Biologists have learned a lot about how viruses operate by watching those whose preferred targets are bacteria.

A particularly well-known bacteria-attacking virus is the one biologists call T2. Like most viruses, T2 is simple. It has a strand of DNA—the molecule that carries hereditary information—enclosed in a coat of protein that looks like a dome-shaped hollow shell with a tube coming out of one side. When a T2 virus attacks a bacterial cell, it uses that tube to punch through the surface of the cell and inject its DNA.

Once injected into the bacterial cell, the virus's DNA takes charge of the cell's chemical processes. Raw materials in the cell are no longer used for the bacteria's own purposes, but are diverted to build new viruses. Around twenty-five minutes after the initial virus attack, the disabled bacterial cell breaks open and releases about a hundred brand-new T2 viruses.

Not all viruses operate exactly like T2. Some spend a long time inside a cell doing nothing—a latent period—until the copying process is triggered by something outside. This is true of a familiar virus that attacks not bacteria but human beings. The herpes simplex virus that causes cold sores in humans has a latent period that can last weeks or months.

Different viruses attack different kinds of cells, and take different amounts of time to act. But no virus can reproduce except by taking over a living cell and using the materials inside to its own advantage.

André Lwoff, "Interaction among Virus, Cell, and Organism," Nobel Lecture, 1965, in *Nobel Lectures in Molecular Biology, 1933–1975* (New York: Elsevier–North Holland, 1977).

THE BIRTH OF THE ATOMIC AGE

In 1895, physicists were excited about a recent discovery: X-rays. Henri Becquerel of Paris wondered whether X-rays would be emitted by certain substances when they were exposed to sunlight.

Becquerel experimented with crystals of a salt containing uranium. He wrapped a photographic plate in black paper, put a few grains of uranium salt on the paper, left the whole arrangement in the sun all day, then developed the plate. An image of the uranium salt grains appeared. Becquerel cautiously concluded that uranium salts, exposed to sunlight, gave off rays capable of penetrating black paper.

One day Becquerel went outside with another wrapped photographic plate topped with uranium salts, planning to leave it outside as before. But the sky was partly cloudy, so he took the plate back indoors and put it in a drawer.

After a couple of days, he developed the plate anyway. To his surprise, Becquerel found an image just as strong as those on plates that had lain in the sun. Without sunlight, this uranium salt had made rays that would go through black paper! Nineteenth-century physics couldn't explain it. Becquerel had discovered a new property of matter: radioactivity.

In his 1903 Nobel Prize lecture, Henri Becquerel anticipated the whole era of atomic energy by guessing that "the emission of energy is the result of a slow modification of the atoms of the radioactive substances. Such a modification . . . could certainly release energy in sufficiently large quantities to produce the observed effects, without the changes in matter being large enough to be detectable by our methods of investigation."

Henri Becquerel, lecture quoted in *Moments of Discovery: The Development of Modern Science*, ed. George Schwartz and Philip W. Bishop (New York: Basic Books, 1958).
"On Radioactivity, a New Property of Matter," Nobel Lecture, 1903, in *Nobel Lectures: Physics, 1901–1921* (New York: Elsevier, 1967).

Honest Answers to Personal Questions

A stranger knocks on your door and says he's taking a poll about tax-law compliance. He asks you whether you've ever cheated on your taxes.

Suppose, for the sake of argument, that you have. You'll probably say no, especially if you have any suspicion that this poll-taker actually works for the IRS.

But suppose the poll is legitimate—the pollsters don't care about your taxes in particular; they actually want to find out what percentage of people cheat on their taxes. You, as a citizen, see value in finding this out, but you don't want your privacy compromised.

A solution is this: the poll-taker gives you a spinner, like the ones that come with children's board games. The circle on the spinner is divided into four equal parts. One of the parts is marked "yes"; the other three parts are marked "no." The poll-taker asks, "Have you ever cheated on your taxes? Don't tell me the answer," he says. "Spin the spinner where I can't see it and just tell me whether the answer is right or wrong for you."

From your answer—"right" or "wrong"—the poll-taker cannot find out whether you personally cheated. But your answer, along with many others, allows him to calculate what percentage of people cheated on their taxes. For example, if everyone cheated, about one-fourth of the respondents would say the spinner was right; if no one cheated, about three-fourths would say it was right.

Spinners and other randomizing devices provide a way to get honest answers to personal questions without violating anyone's privacy. Randomized surveys have already been used to find out about drugs, abortion, and even shady practices in selling cars. Future applications may include finding out more about the connection between sexual behavior and AIDS.

Gina Kolata, "How to Ask about Sex and Get Honest Answers," *Science* 236:382 (April 24, 1987).

RISK ASSESSMENT

In modern times, we often find ourselves evaluating risks in order to make decisions about new technologies and new chemicals. Estimates of risk come to us—from scientists, engineers, and doctors—in the form of statistics that are sometimes hard to grasp, even for experts.

Specialists in the relatively new field of risk assessment offer some suggestions about judging statistics:

We can be critical about data. Is a risk assessment based on historical statistics? Laboratory experiments? How closely do those situations resemble our own?

We can observe how a risk is expressed. Saying there's a .1 percent chance of catastrophe in a certain situation may have a different psychological effect from saying there's a 99.9 percent chance nothing will go wrong.

We can compare the new to the old, the unfamiliar to the familiar: flying versus driving, a nuclear accident versus a dam collapse.

We can compare the risks of two alternatives. We already do this when we balance the risk of getting cancer from bacon treated with sodium nitrite against the risk of getting botulism from bacon not treated with it.

Better decisions require good information, presented in as many ways as possible. Comparing the familiar to the unfamiliar is especially helpful. That's the job of scientists.

Once the scientific part is done, the social part begins. We citizens must interpret the risks in terms of our own values. Sometimes we have to search for the right balance between the interest of the individual and the interest of society. Then we must judge, decide, act, and live with the consequences.

See articles on risk assessment in *Science* 236 (April 17, 1987).

BROKEN SYMMETRY: COSMETIC OR COSMIC?

Mirrors do funny things to reality. Clocks run counterclockwise. Writing is backward. (See "A Mirror Riddle.")

But processes of nature, in which decisions by living organisms are not involved, look perfectly all right in a mirror. If you look at the bathroom faucet in a mirror, you see water going down the drain, not up to the sky. You can't tell, unless the mirror is dirty, whether you're looking at reality or its reflection. In the mirror, no laws of physics are broken—except in a few esoteric physics experiments of the last thirty years, involving particles smaller than the nucleus of an atom: the so-called neutral K mesons.

Flying through a particle accelerator, neutral K mesons change into other particles which fly off in predictable directions. If you're a physicist familiar with neutral K mesons, you can tell from the directions in which these particles fly off whether you're seeing reality or a mirror reflection. Everything else may look perfectly all right, but the behavior of neutral K mesons will look wrong in a mirror.

This is a mystery. What force, or process, causes this tiny exception to the rule that nature and its mirror image are essentially indistinguishable? Physicists have come to suspect that this flaw in the symmetry of nature may be not just cosmetic, but cosmic.

Richard P. Feynman, *The Feynman Lectures on Physics* (Reading, Mass.: Addison-Wesley, 1964), vol. I, chap. 52.

Robert K. Adair, "A Flaw in a Universal Mirror," *Scientific American*, February 1988.

#

Antimatter is not just the stuff of science fiction; it's real. Antimatter is made of particles complementary, in a way, to matter particles. For instance, the electron, a familiar matter particle, has a negative electrical charge. Its corresponding antiparticle, the positron, has a positive charge. When an electron and a positron meet, they annihilate each other. Both particles disappear in a flash of light—a spectacular case of matter changing to energy. The same thing happens in any encounter of an equal amount of matter and antimatter.

Antiparticles are created in nuclear reactions in stars and in space, but they soon meet matter particles and annihilate. So you're unlikely ever to see antimatter, even in a museum.

A great mystery about antimatter was discovered by the physicist Paul Dirac in 1930. Dirac found that, according to the known laws of physics, antimatter has just as much right to exist as matter. So why does so little antimatter exist today?

One guess says that the very early universe—15 or 20 billion years ago—was made of almost exactly equal amounts of matter and antimatter. But, for some reason, there was just slightly more matter—by about one part in a billion. Annihilation eliminated most of it. But some unannihilated antimatter was left over to become our universe.

Whatever force or process it was that upset the symmetry between matter and antimatter in the early universe may still be at work today. That force or process may be the cause of unexpected behavior among subatomic particles like K mesons. (See "Broken Symmetry.")

Clues about the whole universe in the distant past may be lurking among tiny particles in the present.

Richard P. Feynman, *The Feynman Lectures on Physics* (Reading, Mass.: Addison-Wesley, 1964), vol. I, chap. 52.

Robert K. Adair, "A Flaw in a Universal Mirror," *Scientific American*, February 1988.

THE SECRET OF CLEAR ICE CUBES

The problem: cloudy ice cubes, with unsightly bubbles in the center, even though you started with clear water and a clean ice tray.

The answer: start with hot water, not cold.

The reason: hot water holds less dissolved air than cold water. Those bubbles in the center of an ice cube come from air dissolved in the water.

Bubbles usually form at the center because ice cubes usually freeze from the outside. The top, bottom, and sides of the cube freeze first, leaving a liquid water center. As the cube continues to freeze, dissolved air is forced into the liquid center. Air can't freeze at these temperatures, so when the liquid center of the ice cube finally freezes, the air comes out of solution and forms bubbles in the ice. Hot water has less dissolved air to begin with, so it makes fewer bubbles when it freezes.

To convince yourself that hot water holds less dissolved air than cold water, think of what happens when you heat water in a saucepan on the stove. Long before the water gets hot enough to boil, tiny bubbles form on the bottom of the pan. Those tiny bubbles are air coming out of solution as the water warms up. The same thing happens in your water heater.

Or, think of an aquarium: if the temperature is too warm, fish die—partly because the warm water holds too little oxygen.

Getting back to ice cubes: if some dissolved air has already been removed from water by heating, less air will be left to emerge as bubbles when you freeze the water in an ice tray. So, the secret to making clear ice cubes is to start with hot water.

Ronald A. Delorenzo, *Problem Solving in General Chemistry* (Lexington, Mass.: D. C. Heath, 1981), p. 240.

BROKEN CUPS AND ATOMS

You can gather the pieces of a broken coffee cup and fit them together, but they won't stick. The fit may be good, but you can't make it good enough.

"Good enough" means getting the pieces so close that atoms interact.

Atoms separated by more than a few times their own diameter won't interact—they're basically indifferent to each other. The reason has to do with the inner structure of atoms, which is apparent only at very close range.

Every atom has a nucleus with a positive electrical charge surrounded by electrons with a negative electrical charge. Electrical opposites attract, so the nucleus and electrons attract each other.

Seen at a distance, though, an atom shows no obvious sign of positively and negatively charged parts. At a distance, electrical effects of the nucleus and electrons are canceled out. Only at close range—less than a few times the diameter of one atom—do the nucleus and electrons have distinct electrical effects.

In the same way, at a distance of a quarter-mile, a red-and-white-check tablecloth shows no obvious sign of colored squares—it looks pink. Only at close range do the red and white squares look distinct.

If two atoms are brought close enough—in the ballpark of a ten-millionth of an inch—the nucleus of one may attract electrons of the other. The atoms interact; a bond forms.

When a coffee cup breaks, the atoms are pulled apart so their relationship changes from interaction and bonding to indifference. The inner structure of an atom on one side is no longer apparent to an atom on the other side.

If you want the broken pieces of the cup to stick, you will have to fill the gap with something that will get close enough to the atoms on each piece to interact with them: glue.

Rodney Cotterill, *The Cambridge Guide to the Material World* (New York: Cambridge University Press, 1985).

TRAPDOOR CODES

You're going on a trip, and you pack all the items you need into one suitcase. You're very clever in arranging everything so that it just fits. The first night of your trip, you unpack. But the next morning you can't remember how you got all that stuff into the suitcase. Unpacking was easy; repacking is hard, because you have to reconstruct that perfect arrangement that allowed everything to fit in the suitcase.

43

Mathematicians have discovered new ways of transforming messages that work in a similar way. With these mathematical procedures, changing a secret message into code is very easy. But changing the coded message back into readable form is nearly impossible unless you know the secret decoding program.

These are called trapdoor codes, after the one-way trapdoors in mystery stories and horror movies. Going through the trapdoor one way is easy, but you can't go the other way unless you know where the secret button is.

What's new about trapdoor codes is that you need one key to put the message into code, and a different key to translate the message out of code. It's perfectly safe for anyone to know the key for coding a message—because that key won't work for decoding the same message.

To send a message by trapdoor code is easy, like unpacking a suitcase. But to read the coded message is hard, like repacking the suitcase.

Trapdoor codes are already being used by some banks. Suppose you want to send an electronic message in confidence to a bank that doesn't know you. You put the message into the bank's code using a key that the bank publishes in the phone book. But only the bank has the decoding key, so only the bank can read your message. Furthermore, you can use your personal trapdoor decoding key to put a signature in the message, proving that you sent it.

Martin E. Hellman, "Mathematics of Public-Key Cryptography," *Scientific American*, August 1979.

Martin Gardner's "Mathematical Games" column in *Scientific American*, August 1977.

LATE-NIGHT RADIO

Why do distant AM radio stations come in better at night?

The story begins in the rarefied atmosphere high above our heads, where ultraviolet light and X-rays from the sun strip electrons off atoms. The result is a gas made partly of electrons and partly of atoms from which electrons have been stripped. The stripped atoms are known as ions, and the part of the atmosphere where sunlight makes ions is known as the ionosphere.

When radio waves encounter the ionosphere, interesting things happen; ions and electrons have electrical charges that affect radio waves in many complex ways.

A radio wave of the right frequency, coming up from the ground, will be turned around and sent back down by the upper layers of the ionosphere—a hundred or two hundred miles up. The upper ionosphere acts like a mirror, reflecting radio waves around the curve of the Earth to distant receivers. That's why you can pick up a baseball game from a station hundreds of miles away.

But why is nighttime the right time for long-distance AM radio?

The reason is that during the day, the radio waves don't even get a chance to reach those reflecting layers of the ionosphere. The waves are blocked by a low-altitude layer, about fifty miles up, that exists only in the daytime. As long as the sun is up, this low layer prevents AM radio signals from reaching the upper ionosphere. But the low-altitude layer of the ionosphere can't exist without continual sunlight. As soon as the sun goes down, the ions and electrons in this low layer get back together to form ordinary air. The obstruction disappears, and radio waves have a clear shot at the upper atmosphere, a hundred miles up in the night sky.

See entries under "Radio," "Radio Broadcasting," "Ionosphere," etc., in *McGraw-Hill Encyclopedia of Science and Technology* (New York: McGraw-Hill, 1987) and *Encyclopaedia Britannica.*

DOES NATURE ABHOR A VACUUM?

"Nature abhors a vacuum"—or does it?

Abhorrence of a vacuum once had the status of a scientific principle. Then along came seventeenth-century Italian physicist Evangelista Torricelli. In his most famous experiment, Torricelli got skilled Italian glassblowers to make a long glass tube, closed at one end—basically a four-foot-long test tube. He filled the tube with mercury, put his finger over the open end, turned the tube over, immersed the open end in a bowl of mercury, and removed his finger. The mercury in the tube didn't all run out; it fell to about thirty inches above the bowl and stopped. Between the sealed top end of the tube and the top end of the mercury was empty space—a vacuum.

The old idea was that this vacuum pulled the mercury and held it up in the tube. Torricelli tested that idea by repeating his experiment with another tube that had a big bulb at the sealed end. The mercury fell to the same level as before, even though there was now more vacuum at the top of the tube. Torricelli concluded that the vacuum in the tube was not pulling the mercury from above; instead, the weight of the atmosphere was pushing the mercury from below—by pushing down on the mercury in the bowl.

We would now call Torricelli's device a mercury barometer. We still express atmospheric pressure in inches of mercury.

It's possible to build a water barometer on the same principle. But since water is less dense than mercury, atmospheric pressure will push water higher—up to a maximum of about thirty-four feet.

Based on a letter by Torricelli, translated in the anthology *Moments of Discovery,* ed. George Schwartz and Philip W. Bishop (New York: Basic Books, 1958).

How does water get to the top of a tree?

After water enters a tree through the roots, it rises through vessels in the tree all the way to leaves at the top. Some trees are as much as three hundred feet tall. How does water climb three hundred feet? An explanation that's pretty widely accepted among botanists is that the process is driven by evaporation from the leaves.

Whenever a water molecule evaporates from the end of a vein in a leaf, that departing molecule pulls a train of other water molecules lined up in the vein behind it. Those water molecules, in turn, pull the ones behind them, and so on, through the twig and the branch, down the trunk, all the way to the ground, where water is pulled from the soil into the roots. It's as if there were an unbroken thread of water extending through the tree, being pulled at the top end by evaporation.

It sounds strange—this idea that water can be pulled like a thread. But it's possible because the water is confined in a strong airtight tube, like one of those vessels. So this thread of water doesn't break because water molecules have a strong attraction for each other.

We see that same attraction at work whenever we spill water on the kitchen table. The water pulls itself together into drops—it doesn't just scatter and disappear.

So, because of the mutual attraction of water molecules, evaporation from leaves pulls water out the top of a tree. And that, in turn, pulls fresh water—and nutrients—in through the roots. Water gets to the top of the tree because it's pulled up by evaporation.

Frank B. Salisbury and Cleon W. Ross, *Plant Physiology*, 3rd ed. (Belmont, Calif.: Wadsworth, 1985).

Richard P. Feynman, *The Feynman Lectures on Physics* (Reading, Mass.: Addison-Wesley, 1964), vol. I, chap. 1.

ADDING AND SUBTRACTING COLORS

Red plus blue plus green makes white when you mix light, and black when you mix paint. Mixing colored light and mixing colored paint are different processes.

You can see mixing of colored light in a color television picture. All the colors we see in the picture from a distance are made of glowing dots of red, green, and blue light on the screen. These dots glow with different intensities to make different colors. If only the red dots glow, the TV picture is red. If the red dots and the green dots glow, the picture looks yellow, because red light added to green light makes yellow light. If all the colored dots glow at full strength, the picture looks white. Other colors are made by adding the three basic colors in various proportions.

Paint, on the other hand, doesn't make light. It absorbs light. Take the example of a red car parked in the sun. Sunlight already contains all the colors of the rainbow. The red paint on the car absorbs non-red colors. In other words, red paint subtracts just about every color from white light, so it looks nearly black.

When you mix paint, you're subtracting colors. When you mix light, you're adding colors. So you have to think of adding and subtracting color when you try to guess, for instance, the color of a light-blue wall illuminated by a yellow-orange light bulb.

See "Colour" and related entries in *Encyclopaedia Britannica;* "Color Television" and related entries in *McGraw-Hill Encyclopedia of Science and Technology* (New York: McGraw-Hill, 1987).

WHAT COULD CHANGE EARTH'S CLIMATE?

What could change Earth's climate? Theoretical possibilities are almost endless. Here's an example of a natural climate cycle that would tend to make cold periods of geological time even colder. The key to this theory is the presence of a continent at one of the Earth's poles—namely, Antarctica at the south pole.

A continent can hold fallen snow; an ocean can't. A continent at a pole, like Antarctica, can stay cold enough to retain winter snow through the summer. A continent covered with snow is white; it reflects a lot of sunlight back into space. A snow-white continent tends to cool the climate by reflecting (rather than absorbing) sunlight.

A cooler climate makes still more snow, which accumulates on the polar continent—Antarctica—eventually forming a sheet of ice that could spread beyond the boundaries of that continent. The growing ice sheet makes a bigger white area on the Earth, which reflects even more sunlight. That cools the climate more, causing more snowfall—and so the cycle continues.

This cycle is plausible. But is it really happening? Is it important? No one knows for sure.

This is a cycle that would tend to magnify small changes in climate—making a cold period colder. A snow-white continent tends to cool the climate by reflecting sunlight. But, since the world is not covered with ice, this can't be the only process influencing our climate.

There are other plausible cycles that would tend to counteract small changes in climate, helping to keep it the same.

"Glacial Epoch," in *McGraw-Hill Encyclopedia of Science and Technology* (New York: McGraw-Hill, 1987).

POLARIZED LIGHT AND A QUIET LAKE

Look at the water. What's that dark patch in the reflection of the sky?

It's a polarized-light effect. To see it, you need patience, luck, a clear twilight sky, and a quiet lake. The sun has to be on the horizon—just rising or setting. The water has to be like glass. The sky has to be clear blue; a few small clouds are okay, but no haze.

Stand right at the edge of the water, with the sun on your left or right—not in front of or behind you. Look at the surface of the water about four feet in front of you. Examine the reflection of the blue sky. What you're looking for is a dark patch in the reflected sky that you don't see when you look up at the real sky. If you see it, here's the explanation: polarization of light, a quality of light the human eye is not sensitive to.

We can think of light as a vibration, something like a vibration traveling along a stretched rope. Light that vibrates in some clearly defined direction is said to be polarized.

The blue light of the sky is polarized. When the sun is low to your left or right, as it is in this experiment, the blue light of the sky right in front of you is polarized vertically—the vibrations are up and down.

But the lake doesn't reflect that light very well. Horizontal surfaces in general don't reflect vertically polarized light very well. You see a dark patch in the lake four feet in front of you because at that angle, the lake doesn't get much light it can reflect.

The blue sky supplies vertically polarized light; the lake, being horizontal, can't reflect it—and that's why there's a mysterious dark patch in the reflection of the sky.

Marcel Minnaert, *The Nature of Light and Colour in the Open Air* (New York: Dover, 1954), p. 253.

G. P. Können, *Polarized Light in Nature* (New York: Cambridge University Press, 1985).

ARE FOREST FIRES ALWAYS BAD?

In nature, fire can be beneficial. Some forests and other natural communities are accustomed to occasional fires. Fire is a normal part of the life cycle of the grasslands of the Midwest, the chaparral and ponderosa pine forests of the Southwest, and some pinelands of the South.

Fire stimulates the germination of some seeds. It reduces dead plants to ash, releasing nutrients that dissolve in the next rain and quickly return to the soil. Fire can help animals by clearing dead stems that get in the way of grazing. After a fire, new seeds can be carried in by wind or animals; that leads to greater diversity of vegetation.

Fire rejuvenates some forests. If you see a natural stand of firs or pines in which all the trees are the same age, you can usually assume that the trees grew from a seedbed prepared by fire.

There are disadvantages to preventing forest fires completely. If a forest grows for too many years without a fire, dead and decaying plant matter piles up on the ground. Then if a fire does start, it'll be hotter and more destructive than cooler fires that otherwise would happen more often.

But we're not suggesting that it's okay for human beings to be careless with fire in the forest. Natural fires happen in random locations, usually in summer, the time of lightning storms. A forest or grassland that's accustomed to fire can recover and even benefit from an occasional fire of moderate intensity. Unnatural fires—those caused by people—happen at times and places that may make it impossible for the forest to recover.

Robert Leo Smith, *Ecology and Field Biology,* 3rd ed. (New York: Harper and Row, 1980).

DEATH OF THE DINOSAURS REVISITED

Were the last generation of dinosaurs all of the same sex? Here's another theory about that well-known mystery of life on Earth, the disappearance of the dinosaurs 65 million years ago.

According to this theory, the dinosaurs may have become extinct because small variations in the temperature of the climate would cause their eggs to produce all males or all females. The idea came from observations of modern reptiles. The gender of most turtles, alligators, and crocodiles is determined by the temperature at which the eggs are incubated. There is only a small range of temperatures that will produce both males and females from the same brood of eggs. Eggs cooler than that narrow range produce one sex exclusively; warmer eggs produce the other sex.

Mississippi alligators are a case in point. If their eggs are incubated below about 86 degrees Fahrenheit, the babies are all female; above about 93 degrees, all male. Mother alligators ensure a mix of male and female in the next generation by building some nests on hot levees and others in cooler marshes.

If the ancient dinosaurs were like some of their modern relatives, a small change in climate could have produced a whole generation of dinosaurs of one sex. Obviously, if all the newborn dinosaurs were of the same sex, mating soon would have become impossible. Result: extinction.

Scott F. Gilbert, *Developmental Biology* (Sunderland, Mass.: Sinauer Associates, 1985).
M. W. J. Ferguson and T. Joanen, "Temperature of Egg Incubation Determines Sex in *Alligator mississippiensis*," *Nature* 296:850 (April 29, 1982).

Breaking a coffee cup

Why is it so easy to break a coffee cup if it's already cracked?

To break a coffee cup, you have to pull atoms far enough apart—maybe a millionth of an inch—so they no longer bond to each other.

All along the crack, atoms have already been separated. The point of interest is at the tip of the crack—where it ends in solid material. To make the crack just a little longer, all you have to do is separate a few more atoms.

The important principle here is leverage.

Leverage is the principle that allows you to pull a nail out of hard wood with a crowbar. A small force at the end of a long crowbar becomes an immense force pulling on the nail. The crowbar has the effect of magnifying the force you apply to it.

Usually, a cracked coffee cup breaks because something hits it or presses on it with a force. The two sides of the crack act as levers, transmitting that force to the tip of the crack like little crowbars, pulling atoms apart. As the crack proceeds through the cup, these levers get longer, so they pull atoms apart with even greater force. The longer the crack becomes, the easier it is to make it just a little longer.

To break a coffee cup, you don't really have to pull billions and billions of atoms apart all at once—you only have to separate a few atoms at the tip of the crack, then a few more after that, and a few more after that, and so on all the way to the other side of the cup.

In real life all this happens in a split second as the cup hits the floor. High-speed photographs show that cracks travel through brittle materials like ceramics and glass at thousands of miles per hour.

A split second is all it takes to break a coffee cup.

J. J. Gilman, "Fracture in Solids," *Scientific American*, February 1960.
J. E. Field, "Fracture of Solids," *Physics Teacher* 2:215 (1964).

How Bacteria Resist Antibiotics

Here's how it happens in a typical laboratory situation: the case of intestinal bacteria and the antibiotic drug streptomycin. Intestinal bacteria are a well-known and well-studied type, easy to grow in a laboratory; they'll thrive on simple nutrient broth in a flask. That's one reason biologists use them so often for experiments. Intestinal bacteria also multiply fast; one cell can leave a billion descendants in a day.

Most intestinal bacteria are easy to kill. A tiny dose of the antibiotic drug streptomycin kills all the intestinal bacteria in a flask—all except about one in a billion, that is. That one intestinal bacterium in a billion is different from the others in that it cannot be killed by streptomycin. A spontaneous change—a so-called mutation—has taken place in that bacterial cell.

Under normal conditions this mutation is of no use to the bacterium. But if the environment changes, that mutation becomes an advantage. The mutant cell can survive in streptomycin. If streptomycin is in the flask, all the normal bacteria are killed, but a new culture springs from that one-in-a-billion bacterial cell that happens to be resistant.

The individual bacterial cells don't change during their lifetimes. It's the bacterial population as a whole that adapts to changing conditions.

So, bacteria can become resistant to an antibiotic because each new generation produces a few different individuals—mutants. Those mutants are a kind of insurance for the future of the whole bacterial population. If conditions change, maybe one of those different individuals will be able to cope and propagate the bacterial culture.

T. Dobzhansky, "The Genetic Basis of Evolution," *Scientific American*, January 1950; reprinted in *Facets of Genetics: Readings from Scientific American*, ed. A. Srb (San Francisco: W. H. Freeman, 1970).

COOL WIND BEFORE A THUNDERSTORM

A blast of cool air from a threatening cloud signifies that the engine that runs a thunderstorm has started. The simplest isolated thunderstorm begins with hot afternoon sunshine baking the ground. Air near the ground becomes warm, moist, buoyant.

At some point, a mass of this buoyant air cuts loose from ground level and rises to create a warm, moist updraft. That warm, moist air expands in the low atmospheric pressure thousands of feet up. Expansion cools the air, just as it does when air escapes from a pressure nozzle; the moisture condenses. In early afternoon that condensation makes fluffy cumulus clouds; later, it makes towering storm clouds with ice crystals and tiny raindrops.

After a while, raindrops grow big and heavy enough to fall through the updraft, dragging cool high-altitude air down with them to make a cool downdraft in the cloud. Some raindrops evaporate as they fall. That takes heat from the air and makes the downdraft even cooler and stronger. Soon, cool air pours from the base of the cloud and makes a gusty wind ahead of the storm. This is an indication that, somewhere in the cloud, rain has begun to fall—along with snow and ice higher up. The thunderstorm's engine is running: warm air and moisture going up, cool air and rain coming down.

Somehow that engine also electrifies the storm cloud to make lightning, which in turn makes thunder.

After fifteen or twenty minutes, the cool downdraft has grown to cover the whole base of the storm cloud. That cuts off the supply of warm air and moisture from the cloud, and the storm dies.

"Thunderstorm," in *Encyclopaedia Britannica*, 15th ed.
C. Donald Ahrens, *Meteorology Today*, 2nd ed. (St. Paul: West, 1985).
Richard P. Feynman, *The Feynman Lectures on Physics* (Reading, Mass.: Addison-Wesley, 1964), vol. II, chap. 9.

LIGHTNING

When lightning strikes the ground, there's an opening act followed by the main event.

The opening act is a stream of negative electrical charge coming down from the storm cloud in steps. This so-called step leader travels about a hundred and fifty feet at about a sixth of the speed of light, then stops for about fifty-millionths of a second; it then moves on another hundred and fifty feet and stops—and so on toward the ground.

That step leader establishes an electrical connection, a path of least resistance, between cloud and ground. The step leader sets the stage for the main event, the bright flash we see as lightning. That bright flash is the "return stroke": positive electrical charge traveling up from ground to cloud at about a million miles an hour along the path established by the step leader.

That's your basic lightning flash: a faint step leader going down, then a bright return stroke going up, all in a tiny fraction of a second.

The leader-and-stroke cycle usually repeats several times along the same channel. High-speed cameras have recorded dozens of individual lightning discharges along the same path in less than a second.

Plenty about lightning is still unknown. For instance, how does the storm cloud get electrified in the first place? Cloud and ground are like terminals of a ten-million-volt battery. What charges the battery in the first place? Updrafts, downdrafts, ice crystals, and raindrops are probably responsible—somehow. Electrification of storm clouds is one of the unsolved mysteries of weather.

C. Donald Ahrens, *Meteorology Today*, 2nd ed. (St. Paul: West, 1985).
Richard P. Feynman, *The Feynman Lectures on Physics* (Reading, Mass.: Addison-Wesley, 1964), vol. II, chap. 9.

THUNDER

What can the sound of thunder tell you?

Lightning is quick; thunder is slow. The electrical discharge of lightning takes only a fraction of a second, even if the lightning stroke is many miles long. But thunder from that lightning stroke, like other sounds, takes about five seconds to go just one mile.

Another interesting fact about sound in air is that low-pitched sounds travel farther than high-pitched sounds before they die out.

With those ideas in mind, you can interpret thunder.

After a lightning stroke, the first thunder you hear will be from the part of the lightning stroke nearest you. Thunder from more distant parts of the same stroke reaches you later. Lightning is almost instantaneous; thunder is spread out in time.

Thunder from nearby lightning still contains most of its original mix of high and low pitches, so it sounds like ripping or cracking. Thunder from far away has lost its high pitches. What's left is a deep rumble—low pitches of the original thunder, mixed with echoes from the ground.

Lightning may have branches in addition to its main channel, like branches from the trunk of a tree. A ripping sound from nearby branches reaches you first; then, a loud crack from the main channel; finally, a deep rumble from distant branches.

So thunder can tell you something about the shape of the lightning stroke that caused it. Scientists have used stereo recordings of thunder to deduce the shape of lightning strokes hidden in clouds.

While you're listening to thunder, please don't get hit by lightning. Don't stand in the middle of an open field; don't sit under the apple tree; bring the sailboat to shore; get inside a car or a building. Survive to listen to another thunderstorm.

Arthur A. Few, "Thunder," *Scientific American*, July 1975.

Heat Lightning

One of the more mysterious pleasures of a warm summer evening is the spectacle of lightning from distant thunderstorms, flickering silently on the horizon while stars shine overhead. People usually call it heat lightning.

Lightning is easy to see at great distances, especially when it illuminates high, thin clouds visible for many miles. But thunder usually doesn't carry more than about ten or fifteen miles from the storm, because turbulent air around a storm acts as a damper on sound waves.

Another reason thunder doesn't carry very far has to do with differences in temperature between air at ground level and higher up. Early on a summer evening, the ground is still warm from afternoon sunshine, so the air at ground level is also warm. A few thousand feet up, the air is cooler. This temperature difference bends the sound of thunder upward. Here's how it happens.

We can think of a sound wave as an invisible wall of slightly compressed air, traveling at the speed of sound. Sound travels slightly faster in warm air than in cool air. So the part of that invisible wall down in the warm air travels a little faster than the part in the cool air higher up. The bottom of this invisible wall gets ahead of the top as it travels. The invisible wall of sound bends upward as it goes.

The common term "heat lightning" actually describes an essential feature of the situation. Because of a layer of warm air near the ground, the sound of thunder is bent upward, into the night sky. The result is that you see lightning on the horizon but you don't hear thunder.

C. Donald Ahrens, *Meteorology Today,* 2nd ed. (St. Paul: West, 1985).

How many girls, how many boys?

Suppose someone marries and has four children. How many will be girls and how many boys? In this example we're neglecting the possibility of twins. If each of the four children has an equal chance of being a boy or a girl, what combination of boys and girls is most likely?

It seems obvious that the most likely combination would be two boys and two girls. But it's not. To find the answer, all you have to do is list all the possibilities.

There could be a girl, then a boy, then two girls; a boy, then two girls, then a boy; four girls and no boys, and so on. There are a total of sixteen possible combinations.

Out of those sixteen combinations, eight have three children of one sex and one of the other. Only six of the sixteen combinations involve two girls and two boys. So, if a couple has four children, it's most likely that they will have three children of one sex and one of the other. If each child has an equal chance of being male or female, and if there are no twins, that outcome is more likely than two boys and two girls.

You can check your list of the sixteen possibilities and also see that there are only two in which all the children are of the same sex; in other words, there's a two-out-of-sixteen or one-out-of-eight chance of having four children of the same sex.

Martin Gardner, *Entertaining Science Experiments with Everyday Objects* (New York: Dover, 1981).

MODERN BIOLOGY IN A MONASTERY GARDEN

What do you get when you cross a one-foot pea plant with white flowers and a six-foot pea plant with purple flowers? There have always been superstitions and off-color jokes about questions like that. But Gregor Mendel, one of the most important figures in the history of biology, wanted to find out what really happens—by crossing real pea plants.

Mendel lived in a monastery more than a hundred years ago. He experimented with plant breeding in the monastery garden. Mendel had a hunch that heredity proceeded according to rules; he wanted to find the simplest form of those rules. Here's some of what his experiments revealed:

If you cross a one-foot pea plant with a six-foot pea plant, you get plants that are either one foot tall or six feet tall—not in-between, three-and-a-half-foot plants. If you cross purple flowers with white flowers, you get either purple or white flowers—not in-between, lavender flowers. At least not with pea plants.

On top of that, Mendel discovered that if you know the ancestry of the parent plants, you can predict, using a mathematical formula, what percentage of plants in the next generation will have purple flowers as opposed to white.

Actually, heredity is rarely so simple. But Mendel kept his plant-breeding experiments as simple as possible so that he would get clear results. His results showed that some easy-to-see traits like height and flower color were passed from generation to generation in a strict pattern.

Gregor Mendel was a pioneer in what is now called genetics. His garden experiments of over a century ago revealed heredity operating with almost computer-like precision to help make the luxuriant variety of living plants.

G. Mendel, "Experiments in Plant Hybridization" (1865), trans. and reprinted in *Classic Papers in Genetics*, ed. J. A. Peters (Englewood Cliffs, N.J.: Prentice-Hall, 1959), and in

Genetics: Readings from Scientific American, intro. by C. I. Davern (San Francisco: W. H. Freeman, 1981).

SEE YOURSELF AS OTHERS SEE YOU

One mirror is not enough to see yourself as others see you. When you look at a bathroom mirror, you see an image of yourself with left and right reversed. If you don't believe it, extend your right hand to shake hands with yourself. The "person" in the mirror extends his or her left hand.

A bathroom mirror switches left and right in any image it reflects. To see yourself as others do, you need a second mirror to undo the effect of the first mirror and switch the directions back again.

Hold two hand mirrors in front of you with their edges touching and a right angle between them—like the two covers of a book when you're reading. With a little adjustment you can get a complete reflection of your face as others see it. Wink your right eye—the person in the mirror winks his or her right eye. This may seem strange after a lifetime of looking at bathroom mirrors.

When you look at two mirrors held at right angles like the covers of an open book, you see left and right restored to their original relationship. The reason is that the image you see has been reflected twice before reaching you. When you look at the right-hand mirror, you see a reflection of the left-hand mirror, which in turn gives a reflection of the left-hand side of your face, and vice versa. Two reflections are involved.

This seems complicated, but it's easy to see when you try it.

Martin Gardner, *Entertaining Science Experiments with Everyday Objects* (New York: Dover, 1981).

NEUTRINOS AND THE END OF THE UNIVERSE

The universe is expanding; galaxies are getting farther from each other. A big question is whether there's enough matter in the universe to reverse that expansion by mutual gravitational attraction.

Neutrinos are the most abundant and least apparent subatomic particles in the universe. They traditionally have been assumed to exert no gravitational force on anything. But what if that assumption is wrong? If each neutrino exerts even a tiny gravitational force, the combined effect of all of them might reverse the expansion of the universe.

Our understanding of gravity tells us that if neutrinos exert a gravitational force, they must have at least some "rest mass"—the kind of mass that registers as weight on a scale. Do neutrinos have rest mass?

Einstein's special theory of relativity, which has turned out to describe nature accurately so far, says that any object that has rest mass—like a baseball or an airplane—must travel slower than light. So, do neutrinos travel slower than light? We got a chance to check on February 24, 1987, when the explosion of a star—a supernova—was seen in the Large Cloud of Magellan, a nearby galaxy visible from southern latitudes. At about the same time, a burst of neutrinos, presumably from the supernova, reached instruments in Japan and the United States.

Some of the neutrinos arrived at Earth a few seconds later than the others. Was the difference in arrival times due to a difference in departure times? Or were some neutrinos traveling slower—slower than light?

Do some neutrinos have rest mass and exert gravitational force? Will neutrinos reverse the expansion of the universe? The evidence is still ambiguous. We need to know the exact second that light from that supernova first reached Earth. It may or may not be possible to figure that out.

"Supernova Neutrinos," *Scientific American*, June 1987, p. 18.
"Neutrinos from Hell," *Sky and Telescope*, May 1987.

Cooking an Egg

Why does the consistency of an egg change from liquid to more or less solid as it cooks? The important change is in the arrangement of the protein molecules.

A protein molecule is a long chain of smaller molecules, called amino acids. The amino acids are linked by strong bonds between atoms. Those chains are not likely to break while you're cooking an egg. But another change happens when you turn on the heat under a raw egg.

In a raw egg each protein molecule is folded up into a compact ball. There are weak bonds between atoms that hold the protein molecule in its folded-up position. When you heat the egg, you increase the tiny random jiggling motion of the molecules. In any material warmer than absolute zero, the atoms and molecules move around at random. Higher temperature means faster random motion.

As the egg heats, the random motion gets fast enough to break the bonds that keep the proteins folded up. So the protein molecules unfold. The kind of weak bonding that once held the protein molecules in a folded position now works in another way. Here and there, a loose end of one protein molecule comes alongside a loose end of another. The loose ends overlap and bond side to side. As the egg gets hotter, the spliced proteins form a mesh, with water filling in the spaces within the mesh. As more protein molecules unfold and connect to each other, the mesh gets stronger, and the egg becomes more solid. When the mesh is strong enough for your taste, you take the egg off the heat.

So when you cook an egg, the important change is in the arrangement of the protein molecules. They unfold, connect to each other, and form a mesh that gives the egg its new, solid, cooked consistency.

Harold McGee, *On Food and Cooking: The Science and Lore of the Kitchen* (New York: Macmillan, 1985).

How a field can become a forest

The process goes by the name of ecological succession, and in the case of a field becoming a forest, it may take over a hundred years. Here's how it might happen.

Start with an abandoned farm field. The first plants to move in are weeds—the so-called pioneer colonists, including dandelions and ragweed, among others. These pioneers devote a large part of their energy to reproduction.

After two or three years, perennial grasses and shrubs begin to replace the early weeds—that's the next stage in the ecological succession.

In some areas, pine trees move in next, thanks to seeds carried into the field by wind and animals. Within a few decades, what used to be a bare field becomes a pine forest that may last for three-quarters of a century. Earlier grasses and shrubs disappear because they can't tolerate the shade under the pines.

While the pine forest develops, oak and hickory seeds enter the area. Some types of oak and hickory saplings tolerate shade well, so they grow steadily and, after many decades, become tall enough to cast shadows on the pines. Pine trees won't reproduce in shade, so they gradually disappear by attrition due to storms, disease, or old age. The oaks and hickories come to dominate the forest for a nearly indefinite number of decades, or until they are removed by some disturbance. The field has become a forest in the so-called climax stage of ecological succession.

This is ecological succession in a simple idealized example. The specifics vary from region to region, and in any case few fields remain absolutely undisturbed for a hundred years or more.

Eugene P. Odum, *Basic Ecology* (Philadelphia: Sanders College, 1983).

Robert Leo Smith, *Ecology and Field Biology*, 3rd ed. (New York: Harper and Row, 1980).

THE MOST IMPORTANT FLY IN THE HISTORY OF SCIENCE

This fly emerged in 1910—in a bottle, in Thomas Hunt Morgan's laboratory at Columbia University in New York. It was a male fruit fly with a very unusual, unexpected trait: white eyes. Fruit flies usually have bright red eyes.

Morgan was breeding fruit flies to observe patterns of heredity. Wondering which of the descendants of this unusual white-eyed fly would inherit white eyes, he bred it to its red-eyed sisters. Two fruit-fly generations later, some more white-eyed flies appeared.

The strange and surprising thing was that the new white-eyed flies were all males. Why did only male flies inherit white eyes? It can be explained in terms of chromosomes—objects in the nucleus of each cell that look like little pieces of thread under a microscope.

White eyes in a fruit fly are caused by an unusual change, an apparent defect, in the so-called X chromosome. A male fruit fly has only one X chromosome in each cell; if that chromosome is defective, the fly has white eyes. A female fruit fly, on the other hand, has two X chromosomes; if one is defective, the other serves as a backup and the female fly has normal red eyes.

Back in 1910, biologists already knew that male fruit flies have a distinctive set of chromosomes. Male fruit flies also have a distinctive capacity to inherit white eyes—that was Morgan's discovery.

Thomas Hunt Morgan's observations gave a clue that the mystery of biological inheritance might be explained in terms of chromosomes—a new idea in 1910. Now it's a basic idea in our understanding of how living things reproduce and develop.

T. H. Morgan, "Sex Limited Inheritance in *Drosophila*" (1910), reprinted in J. A. Peters, ed., *Classic Papers in Genetics* (Englewood Cliffs, N.J.: Prentice-Hall, 1959); originally in *Science* 32:120–122.
Helena Curtis, *Biology,* 4th ed. (New York: Worth, 1983).

How plants fight

For centuries people have known that few plants will grow near a black walnut tree. The tree seems to do something to the soil that inhibits the growth of other plants.

This is more than just competition. In competition, plants vie for the same resources—minerals, water, space, light. Each competing plant in effect seeks to take something from the environment, and to take more than surrounding plants. Black walnut trees use the strategy of putting something into the environment that makes life difficult or impossible for other plants. Tomatoes, for example, wilt and die near black walnut trees; alders won't survive more than about ten years.

The black walnut's weapon is a substance botanists call juglone. It's harmless within the tree, but it's chemically changed into its effective form when it mixes with ground water. Juglone added to plant beds stunts or kills the plants.

In the desert, where resources are especially precious, shrubs like sagebrush, wormwood, and salvia eliminate competition with a family of chemicals called terpenes. Desert shrubs are often surrounded by a zone practically bare of vegetation; even after a fire, new plants have trouble growing in the terpene-affected area.

Some plants even produce chemicals that inhibit their own reproduction. In the prairies of Oklahoma and Kansas, some of the so-called pioneer weeds—sunflower, ragweed, crabgrass, and others—make chemicals that inhibit the growth of other individuals of the same species, cutting out competition even from their own kind.

"Allelopathy," in *McGraw-Hill Encyclopedia of Science and Technology* (New York: McGraw-Hill, 1987) and *Yearbook* (1988).
A. Sutton and M. Sutton, *Eastern Forests* (New York: Knopf, 1985).

How Fast Are The Clouds Moving?

Go outside and glance at the clouds. Are they moving? In which direction? What's the slowest apparent motion your eyes can detect? How quickly can you detect the motion? How does your ability change under different conditions—high and low clouds, strong and light wind, day, night, moon, no moon, overcast, partly clear?

We detect motion either by comparing the position of the cloud to the position of some fixed object like a chimney, or by using our own eyes as the fixed reference. Of course, our eyes don't measure the speed of a cloud in miles per hour; we judge motion in terms of angles within our field of vision.

Our visual perceptual system is a surprisingly sensitive detector of motion. At night, an airplane or a satellite stands out among the stars purely because of its motion, even among stars that are brighter. Some people can detect motion as slow as one-thirtieth of a degree of angle per second, using a fixed object as a reference. At that speed a cloud takes about fifteen seconds to cross the disk of the sun or moon. (The sun and moon, incidentally, happen to have almost exactly the same apparent size in the sky—about one-half a degree of angle.)

Not only are we visually sensitive to slow motion, but we can judge the direction and speed of motion quickly. A brief glance may be enough to tell in which direction and how fast the clouds are moving. A good baseball outfielder can judge the direction of a fly ball within a fraction of a second after it leaves the bat.

Apparently we don't perceive motion by taking one visual snapshot now, another one later, and comparing the two. Instead, we perceive motion well because our eyes are especially sensitive to any change in the pattern of light falling on our retinas.

Marcel Minnaert, *The Nature of Light and Colour in the Open Air* (New York: Dover, 1954).

G. Johannson, "Visual Motion Perception," *Scientific American*, June 1975.

S. H. Bartley, *Introduction to Perception* (New York: Harper and Row, 1969).

K. von Fieandt and I. K. Moustgaard, *The Perceptual World* (New York: Academic Press, 1977).

Sort nuts by shaking the can

If you shake a can of mixed nuts for a few seconds, the largest nuts come to the top. The spaces between nuts are not big enough for small nuts to fall through, but the small nuts end up on the bottom and the large ones on top anyway. Shaking creates momentary gaps in the mixture; small gaps occur more often than large ones.

Then the nuts are sorted by gravity. As the can shakes, the large nuts frequently move aside far enough to allow a small nut to fall into the space beneath. This happens much more often than the reverse process, in which several small nuts happen to make a gap that one large nut can fall into.

Every time a large nut moves far enough to allow a smaller nut to fall into a gap beneath it, that large nut ends up resting on top of the smaller nut. Over the course of several seconds of shaking, the large nuts slowly move up.

This is not only an amusing kitchen observation, but something with practical usefulness. In many parts of the world, construction workers separate gravel from sand by shaking the container. Coarse gravel comes up to the top. Manufacturers can exploit the tendency of large particles to move up when they need to make a mixture of particles of different sizes, in pharmaceuticals, glass making, and paint making. They can put large particles into a container first, then add smaller particles on top. Shaking the container for the right amount of time causes the large particles to move up until they're evenly distributed through the mixture.

"Nuts and Jolts," *Scientific American*, May 1987, p. 58D.

A. Rosato et al., "Why the Brazil Nuts Are on Top: Size Segregation of Particulate Matter by Shaking," *Physical Review Letters* 58:1038–1040 (March 9, 1987). Of related interest: R. B. Prigo, "Liquid Beans," *The Physics Teacher*, February 1988, p. 101.

Everybody Talks About Genes— But What Do They Do?

The classic experiments in the history of science are often memorable not for complexity but for cleverness. Here's an example.

By the 1940s, biologists were pretty well convinced that offspring grow to resemble their parents because information of some kind is transmitted from generation to generation by the so-called genes. But what information is transmitted? Exactly what does a gene do?

The strategy of the biologists George Beadle and Edward Tatum was to "find out what genes do by making them defective." Beadle and Tatum worked in the 1940s with ordinary bread mold. They grew mold in test tubes with a simple nutrient formula: sugar, minerals, and one vitamin. Normal bread mold could make all the proteins and other substances it needed from that nutrient formula.

Then Beadle and Tatum exposed some of their mold samples to intense X-rays. A few of the X-rayed molds would no longer grow unless they had proteins added to their diet. The X-rays had destroyed the molds' ability to make those proteins.

What's more, this inability to make certain proteins, like any other inherited trait, was passed on to future generations of molds. The X-rays had caused a defect, and the defect was inherited. The real effect of the X-rays had been to damage the molds' genes.

The Beadle-Tatum experiment was important because it showed that what a "gene" conveys from generation to generation is a set of recipes—instructions—for making proteins. The next project in biology, still unfinished, is to find out how proteins make offspring resemble their parents.

G. Beadle, "The Genes of Men and Molds," *Scientific American*, September 1948; reprinted in *Genetics: Readings from Scientific American*, intro. by C. I. Davern (San Francisco: W. H. Freeman, 1981).

Looking High and Low at Leaves

Have you ever noticed that the same tree may have leaves of different shapes?

Next time you have a chance, look closely at the leaves on a big tree—a white oak, for example. Compare leaves at the top of the tree with leaves near the bottom. Leaves near the top often tend to be relatively small and thick, the ones at the bottom large and thin. Also, leaves near the top may have deeper lobes than those near the bottom.

It may be that these differences exist because they help the leaves to do their job better. The most obvious job of a leaf is to make food for the tree from carbon dioxide, water, and sunlight. In view of this, a small, thick leaf with deep lobes seems better suited for the sunny, hot environment near the top of a tree. A thick leaf absorbs more carbon dioxide, while losing less water, than a thin leaf of the same surface area. That's because carbon dioxide is absorbed through special cells inside the leaf; a thicker leaf has more of those special cells, for every square inch of surface area, than a thin leaf does.

Also, small size and deep lobes allow a pattern of air circulation that helps keep the leaf cool near the top of a tree. If a leaf gets too hot, it lets too much water escape into the air by evaporation. And if a tree loses too much water, it can't bring the minerals it needs up from the ground.

On the other hand, a large, thin leaf with a smooth outline seems better suited to the darker, cooler environment near the bottom of a tree. Down there, heat and water loss are not so threatening.

So the hot, sunny environment near the top of a tree seems to favor small, thick leaves; the cool, dark zone at the bottom favors large, thin leaves. On many trees, you can see that difference.

D. F. Parkhurst and O. L. Loucks, "Optimal Leaf Size in Relation to Environment," *Journal of Ecology* 60:505–537 (1972).
Susan Wintsch, "The Greedy Leaf," *Garden*, May–June 1986.

A Taste Test

An old-fashioned English lady comes to you and says: "I like milk in my tea. But I'm very particular; the milk must be added to the tea. I do not like it if the milk goes into the cup first. I can tell, just by tasting, whether the milk or the tea went into the cup first."

You, the scientist, decide to test this remarkable claim with an experiment. You prepare eight cups of tea. In four of them, you put the milk in first; in the other four, you put the tea in first. You don't tell the lady which is which. You ask her to taste all eight cups and select the four in which the milk went in first.

The English lady takes a sip from each cup of tea and sets aside four cups, saying, "These are the ones in which the milk went in first." You check your records and find she was right about three of the cups, and wrong about one.

A reporter is on the phone, asking, "Inquiring listeners want to know—can this lady tell, just by tasting, whether the milk was added first? Yes or no?" You're the scientist. What do you do?

You have to keep a cool head and consider all the possible ways the experiment could have come out. What's the probability that the lady could have achieved that degree of success just by guessing?

There are precise mathematical ways of answering that question. In this example, a little calculating shows that there's a 24 percent chance that the lady could have achieved at least that degree of success just by guessing, without having the ability she claims to have. From that you'd probably conclude that you do not have enough information to tell for sure whether the lady can really distinguish which cups of tea had milk put in first.

This is a cute example with a serious moral: a scientist has to consider not only how an experiment did come out, but also all the other ways it could have come out.

Sir Ronald A. Fisher and Ghiuean T. Prance, *The Design of Experiments*, 9th ed. (New York: Hafner, 1974), chap. II.

WHY A RUBBER BAND SNAPS BACK

Stretch a rubber band; it becomes long and thin. Let it go; it snaps back to its original short, fat shape, ready to be stretched again.

Rubber has this useful property for two reasons. First, rubber molecules have a peculiar structure and arrangement; second, those molecules are always moving around, because the rubber is warmer than a temperature of absolute zero. In any material warmer than absolute zero, the molecules are always moving around in a tiny, random, jiggling motion.

Rubber is made of molecules shaped like strands of spaghetti. If you stretch a rubber band, you pull those spaghetti-shaped molecules into a more or less straight line. But the molecules are still moving around. They shake from side to side and bump into each other. Because of that motion, the molecules tend to spread out sideways; so they must "unstraighten," curl up, kink, tangle. That makes them pull inward on the ends of the rubber band. The stretched rubber tries, so to speak, to become short, thick, and flabby so the molecules will have more room to move around sideways. The rubber band snaps back.

This picture also explains another property of rubber: unlike most materials, it shrinks when you heat it and expands when you cool it. In a warm rubber band, the molecules move faster, tend to shake sideways more, and therefore pull harder at the ends than in a cold rubber band. A rubber band will squeeze a package harder if it's been in the sun than if it's been in the freezer.

Richard P. Feynman, *The Feynman Lectures on Physics* (Reading, Mass.: Addison-Wesley, 1964), vol. I, chap. 44.

Frederick T. Wall, *Chemical Thermodynamics,* 2nd ed. (San Francisco: W. H. Freeman, 1965), chap. 15, "Statistical Thermodynamics of Rubber."

Breaking the Tension

Carefully fill a glass with water, until the surface of the water is exactly level with the brim of the glass. You may want to use a second glass to add the last of the water, rather than trying to do it at the faucet. Now gently drop a quarter into the glass. The water surface will bulge upward slightly, but water will not run over the rim of the glass.

Molecules of water attract each other. That attraction makes a film under tension at the surface of the water. Even though the water is bulging upward, surface tension keeps it from spilling.

Now the game is this: how much change can you drop into the water before water spills over the rim of the glass? Every coin displaces an amount of water equal to its own volume. In other words, the bulge at the top of the water has the same volume as all the coins added to the glass. The film at the surface of the bulging water—caused by attraction of water molecules—holds the water like a bag. How much water can that bag hold before it breaks?

Possibilities for elaboration suggest themselves. You could agree that the next-to-last person to add a coin before the water spills wins all the change in the glass.

To make the game last longer, use paper clips instead of coins. As before, carefully fill the glass with water until the water surface is exactly level with the glass brim. How many paper clips can you drop in before the surface tension breaks and water spills over the edge? Some people may guess that about ten paper clips will do it. But remember, a paper clip is just a piece of bent wire, which displaces only a tiny volume of water. You may be able to get over a hundred paper clips into the glass before the water spills.

Of related interest: Martin Gardner, *Entertaining Science Experiments with Everyday Objects* (New York: Dover, 1981).

How cockroaches get away

The all-too-familiar American cockroach almost seems to know where you're going to strike. What's the tip-off that sends the cockroach running?

Entomologists who have investigated this question have found that cockroaches detect the puff of wind generated by a nearby moving object. They run away from the wind. And it has to be a puff of wind, not a steady breeze. To make the roach run, the wind speed has to increase sharply over a small fraction of a second.

Roaches have special organs for detecting puffs of wind. Extending from the rear end of a cockroach are two tapered appendages, the so-called cerci. The underside of each cercus has about 220 delicate hairs connected to the roach's nervous system. When these hairs are struck by a puff of air, they cause nerve impulses to be sent to the insect's leg muscles.

Each hair can flex a little more easily in one direction than in others, so each hair is especially sensitive to wind from a particular direction. Nerve impulses from the hairs are sent to the roach's legs in just the right pattern to cause the roach to turn away from the direction of the wind. A puff of air from the left causes impulses to be sent to the cockroach's left legs; as a result, the roach turns right, away from the wind.

A cockroach with one cercus damaged or removed makes wrong turns. A roach with both cerci damaged or removed doesn't try to escape at all.

J. M. Camhi, "The Escape System of the Cockroach," *Scientific American*, December 1980.

THE MOON ILLUSION

For thousands of years people have been noticing that the moon looks much bigger when it's very low in the sky than when it's high overhead. This effect has come to be called the moon illusion.

It's not a physical effect; the atmosphere does not magnify the image of the moon. The atmosphere may cause the moon to appear flattened or colored, but not magnified. Photographs show that the moon's image is really the same size no matter how high or low it is in the sky. But people almost always judge the image to be larger when the moon is low.

The real cause of the moon illusion seems to be the juxtaposition of the moon and features on the distant horizon. Our perception of the distance to the horizon influences our judgment of how big the moon is. Professional psychologists have done experiments that indicate this, but you can investigate the moon illusion yourself.

Cut a hole the size of a quarter in a big piece of cardboard. Hold the cardboard at arm's length, not next to your eye. Look at the rising or setting moon through the hole. That cardboard screen blocks your view of the landscape near the moon. Does the moon look smaller with the cardboard than it does without?

Another experiment involves looking at the whole scene upside down, by bending over and looking between your legs. For most people, looking at a scene upside down weakens the impression that the horizon is far away. Does that weakened impression of distance dispel the moon illusion for you?

Looking at the rising or setting moon through a hole in a cardboard screen, and looking at it upside down are two simple tricks that may dispel the moon illusion. Try them, and get a hint of the mystery and complexity of our ability to judge distance and size.

Lloyd Kaufman and Irvin Rock, "The Moon Illusion," *Scientific American*, July 1962.

Polarizing sunglasses

How do polarizing sunglasses cut glare?

To start, an analogy. Two children hold a long jump rope stretched between them. The child at one end shakes the rope, causing waves to travel from one end of it to the other. The child can shake the rope up and down, sideways, or in any other direction. That makes waves that vibrate in different directions.

If we think of the waves traveling along that rope as a model of light waves traveling through space, then the direction in which the rope vibrates corresponds to the direction of the so-called polarization of the light. Light can be polarized horizontally, vertically, or in any other direction.

Back to our rope analogy: imagine now that the children pass the rope between the slats of a picket fence. If the child at one end shakes the rope from side to side, the vibrations will be stopped by the fence. But if the shaking is up and down, the waves will go right through.

Polarizing sunglasses are like that picket fence: they pass light that vibrates vertically but block light that vibrates horizontally. In general, light that has bounced off a horizontal non-metallic surface—like a road or a lake—vibrates horizontally. That's glare, and it's the kind of light polarizing sunglasses are designed to block.

Here's an experiment: take the polarizing glasses off, slowly rotate them to a vertical position while you look through one lens, and you'll see glare spots reappear. You are, in effect, turning the picket fence sideways, so the slats let vibrations of glare get through.

G. P. Können, *Polarized Light in Nature* (New York: Cambridge University Press, 1985).

Pros and Cons of the Mercator Projection

If you sail northeast from Caracas, Venezuela, into the Atlantic Ocean, maintaining a heading of northeast, where will you reach land? Portugal? Spain? France? England? Ireland? Scotland? Norway? It's easy to find out, if you use the right kind of map.

The right kind of map for this purpose is the Mercator projection, first published as a world map in 1569 by the German geographer and engraver Gerardus Mercator. Projection means transferring points from the round Earth onto a flat map. There are dozens of projections in use today, but Mercator's projection has a special property: the course that results from maintaining a steady compass bearing, such as northeast, comes out as a straight line. Such a course is technically called a loxodrome.

A loxodrome is not the shortest course between two points on the Earth, nor is it a straight line on the Earth. But a loxodrome is a straight line on a Mercator-projection map, and it's the simplest course on the Earth if you're navigating by compass, as people did in Mercator's time.

To get these advantages, Mercator had to introduce some famous distortions. Continents and oceans near the poles are much too large compared to features near the equator. Greenland looks much too big compared to Africa. For that reason, modern atlases avoid Mercator's projection for maps showing distribution of people or resources.

We'll leave it to you to find out where a northeast course from Caracas takes you. Just lay a ruler on the map at a forty-five-degree angle, with one end at Caracas. But for this purpose, be sure that your map is based on the Mercator projection.

"Map," in *McGraw-Hill Encyclopedia of Science and Technology* (New York: McGraw-Hill, 1987).

Encyclopaedia Britannica, 14th ed. (1968).

There is a Mercator map of the world in *Chambers's Encyclopedia* (1969), vol. XV, pp. 2–3.

A Surprise at a Shadow's Edge

On a sunny day, look closely at your shadow on a plain, smooth surface. You will see that the edge is not perfectly sharp; there's a gradual transition from the dark inside, through a range of intermediate greys, to the bright outside. Now look for something else: a bright band just outside that grey zone on the bright side. Look more carefully and you may also see a dark band just inside the grey zone on the dark side.

That bright band on the outside of the shadow and the dark band on the inside were first discussed in scientific literature over a century ago, by the Austrian physicist Ernst Mach. They're generally called Mach bands.

The Mach bands are an optical illusion: they don't represent actual variations in the amount of light hitting the surface; the bands are created by our visual system. To see them more clearly, try moving around and watching the edge of the moving shadow.

The Mach bands indicate that our visual system accentuates contrast at edges. Our visual system tells us about the difference between the dark area and the bright area by accentuating the line where they meet. That's why we can use lines to draw pictures. To make a convincing picture of, say, a coffee cup, an artist needs to draw only the edges. Our perceptual system interprets the lines to mean "coffee cup" on one side and "background" on the other.

Look for the Mach bands along a shadow's edge: an extra-bright band along the bright side, an extra-dark band along the dark side. Those bands provide a striking piece of evidence that our visual system accentuates contrast wherever we see an edge.

F. Ratliff, "Contour and Contrast," *Scientific American*, June 1972.

Marcel Minnaert, *The Nature of Light and Colour in the Open Air* (New York: Dover, 1954).

Of related interest: M. S. Livingstone, "Art, Illusion and the Visual System," *Scientific American*, January 1988.

Richard P. Feynman, *The Feynman Lectures on Physics* (Reading, Mass.: Addison-Wesley, 1964), vol. I, chap. 36.

Algae as a Thermostat

It has been suggested that algae in the oceans help keep Earth's climate steady. The suggested process goes like this:

Some algae living in the open sea are known to make a gas called dimethyl sulfide, or DMS. This gas escapes from the algae into the air and undergoes a chemical reaction that leaves tiny crystals, about a millionth of an inch across, suspended in the air above the oceans.

Water vapor condenses around these tiny crystals to make cloud droplets—maybe smaller-than-usual cloud droplets. Smaller-than-usual cloud droplets make whiter-than-usual clouds, which reflect more sunlight than usual back into space, cooling the climate.

Now for the more speculative part of this idea. It may be that the algae that make DMS gas prefer warm water. If so, these algae could act as a global thermostat. Here's how.

Suppose the world warmed up for some reason. These particular algae would flourish in the warm water, making more DMS, which would form more tiny crystals in the air, making more and brighter clouds. The extra clouds would reflect more sunlight into space, cooling the atmosphere and reversing the warming trend that started the whole process. In other words, if the algae get too warm, they cause clouds to form and put the ocean in the shade.

That's the theory. Is it right? Not all the evidence is in yet.

In any case, there are many other processes, including the burning of fossil fuels by humans, that could influence climate. A big job in climatology is to find out which processes are the important ones.

"Climate Control," *Scientific American,* July 1987, p. 24.

"No Longer Willful, Gaia Becomes Respectable," *Science* 240:393–395 (April 22, 1988).

R. J. Charlson et al., "Evidence for the Climatic Role of Marine Biogenic Sulfur," *Nature* 329:319 (April 24, 1987).

BALANCE A YARDSTICK WITHOUT LOOKING

A yardstick is thirty-six inches long, so you know the exact center is at the eighteen-inch mark. But how could you find the center blindfolded?

You might guess something complicated, like using the distance between two knuckles to represent one inch, and measuring with that. But there's an easier way. Hold your hands in front of you, with your index fingers pointing forward. Rest the yardstick across your fingers and slowly bring your hands together under the stick. Your fingers will meet under the eighteen-inch mark.

You're actually finding the center of gravity of the stick, the point where it balances. We're assuming that your yardstick is made of wood or other material that's uniform from end to end, with no metal attachments or big holes that would cause it to balance somewhere other than the center. If that assumption is correct, the yardstick will balance at the eighteen-inch mark.

As you bring your hands together, you feel the stick sliding first over one finger, then over the other. Whenever your left hand gets closer to the center of gravity than your right hand, the yardstick presses down harder on your left index finger. More pressure causes more friction, so the left hand stops sliding. All the sliding then happens over your right index finger till your right hand gets closer to the center and reverses the situation. In other words, if one hand gets ahead, friction stops it from sliding till the other hand catches up.

All this happens with no special effort on your part. Just rest the yardstick across your index fingers and slowly, steadily, bring your hands together. Your hands will meet at the eighteen-inch mark, the center of gravity of the yardstick.

H. Steinhaus, *Mathematical Snapshots* (New York: Oxford University Press, 1950).

Martin Gardner, *Entertaining Science Experiments with Everyday Objects* (New York: Dover, 1981).

A ROCK IN A ROWBOAT

Here's a classic puzzle you'll enjoy pondering.

You're in a rowboat in a small pond. You have a huge boulder with you in the boat. You throw the boulder overboard into the water. Does the water level of the pond rise, fall, or stay the same?

Now, it's pretty clear that if you took everything completely out of the pond—yourself, the boat, and the boulder—the water level would fall. And if you stayed in the boat, in the water, and threw the boulder ashore, the water level in the pond would also fall. Since the rowboat would weigh less after you tossed the boulder ashore, the boat would float higher and displace less water. But what happens if you toss the boulder overboard, into the water?

You can simulate the situation by filling a big mayonnaise jar with water. That's the pond. Put a rock in a tin can—that's the rowboat—and float it in the water in the jar. Mark the level of water in the mayonnaise jar. Next take the rock out of the can and drop it into the jar. Now you have a rock at the bottom and the empty tin can floating at the surface. Don't let it capsize. Is the level of water in the jar higher, lower, or the same as before?

Here's the logic behind the answer. When the boulder is in the boat, it displaces an amount of water equal to its weight. That's a lot of water. Lying at the bottom of the pond, the boulder displaces only an amount of water equal to its volume. That's less water. The bolder displaces less water if it's on the bottom of the pond than if it's in the boat. Also, the boat floats higher without the rock.

So here's the answer: if you're in a rowboat with a big boulder and you throw the boulder overboard into the water, the pond water level falls.

R. J. Brown, *333 More Science Tricks and Experiments* (Blue Ridge Summit, Pa.: Tab, 1981).

Whiter and Brighter Than New

The main purpose of laundry detergent is to take dirt out of fabric. But many detergents also leave something behind—an additive designed to make the fabric look whiter and brighter than new.

This additive is a so-called optical brightener. It's really a dye with a special property: it absorbs invisible ultraviolet light from the sun and gives off a faint glow of visible blue light. The blue glow counteracts the yellowish color of dirty old fabric and gives the eye the impression of white.

Optical brighteners are examples of so-called fluorescent materials. These brighteners take in energy from the ultraviolet rays of sunlight and release that energy in the form of blue light visible to the eye.

Optical brighteners can be economical to use. If you add optical brighteners to your clothes, you need less bleach to make them look clean. That can help your clothes last longer, because some bleaches weaken the fibers in the cloth.

Optical brighteners have an especially dramatic effect under black lights, which are still used in some bars and skating rinks and other places to create a festive atmosphere. Black lights produce a lot of ultra-violet light and very little visible light. Fabrics—especially light-colored fabrics—light up as the optical brighteners in the fibers change that invisible ultraviolet light into visible blue light.

"Bleaching," in *McGraw-Hill Encyclopedia of Science and Technology* (New York: McGraw-Hill, 1987).

F. L. Wiseman, *Chemistry in the Modern World* (New York: McGraw-Hill, 1985).

Hormone Insecticides

In human beings, certain hormones trigger the physical changes that make a child's body mature into an adult's body. With insects, it's different. It's the lack of a hormone that allows an insect to mature to its next life stage.

It's called the juvenile hormone, because its function is to keep the insect young—to slow down its development. Being immature for the right amount of time is important for insects. Many do most of their eating before they become adults. After they become adults, these insects mostly travel and reproduce.

Some plants make an anti-juvenile hormone—a substance that destroys the cells in an insect's body that make juvenile hormone. In particular, a bedding plant of the genus *Ageratum* makes an anti-juvenile hormone that disrupts the life cycle of milkweed bugs. Milkweed bug nymphs quickly become miniature adults after exposure to anti-juvenile hormone. That means the bugs eat less during their lifetimes, because they don't get a chance to pass through all their early life stages—the stages in which they would do a lot of their eating. This has obvious advantages for the plant. Female milkweed bugs that have been rushed into adulthood by anti-juvenile hormone are sterile. Also, they refuse to mate.

There has been no large-scale practical application yet, but if anti-juvenile hormones could be manufactured in quantity, they might be used as insecticides that affect only insects. A spraying or dusting of anti-juvenile hormone might cause a whole population of insects to become precocious adults, with a reduced appetite for food and no ability to reproduce.

W. S. Bowers et al., "Discovery of Insect Anti-juvenile Hormones in Plants," *Science* 193:542–547 (August 13, 1976).

Scott F. Gilbert, *Developmental Biology* (Sunderland, Mass.: Sinauer Associates, 1985).

AN INVERTED IMAGE

This is a simple visual experiment. To do it, you need two three-by-five index cards.

Take one of the cards and make a tiny triangle of holes in it with a straight pin. Make the three holes about a sixteenth of an inch apart. Take the other index card and poke just one pinhole in it.

Now take the card with the triangle of pinholes, and hold it very close to your eye. Hold the other card, the one with the single pinhole, about four inches in front of the first one. Face a strong light and look through the three holes at the single hole. You will see your triangle of pinholes upside down.

The holes in the two cards cast three thin beams of light into your eye, making three dots on your retina. Under normal circumstances, your retina gets a focused, upside-down image projected by the lens of your eye. Your nerves and brain in effect turn that image over to make it right side up. But in this experiment, the card with the three holes is so close to your eye that the lens can't focus the three dots into an image. The rest of your visual system, operating as usual, inverts the triangle of light on your retina anyway, and you see the pinhole pattern upside down.

It's a very striking effect; try it.

C. J. Lynde, *Science Experiences with Ten-Cent Store Equipment*, 2nd ed. (Scranton, Pa.: International Textbook Co., 1951), p. 132.

A NEW PERSON FROM A NOSE?

Can you really make a new person from a nose, the way Woody Allen did in the movie *Sleeper*?

First we have to back up. One of the most mind-boggling mysteries of our planet is this: how does a single fertilized egg cell generate the hundreds of cell types in a complete living organism?

Within that question is another question. The fertilized egg divides into two cells, then four cells, then eight, sixteen, and so on, to make all the cells in the new body. Do all those body cells preserve all the genetic information that was in the original fertilized egg? And is that genetic information always available? Would it be possible to grow a new copy of an organism from any of its cells?

It can be done with plants. It appears to have been done with frogs. But with mammals, it's another story.

In one experiment, the nucleus of a fertilized mouse egg cell was replaced with a nucleus taken from another fertilized mouse egg cell. This involves tricky microsurgery. The egg cell with the replacement nucleus, implanted in a female mouse, developed into a normal mouse.

But in another experiment, the replacement nucleus came from a fertilized mouse egg that already had divided into four cells. The egg that received this nucleus did not develop beyond a very early embryonic stage. In other words, cells in a developing mouse embryo almost immediately lose their ability to generate new, separate mice.

The real purpose of these experiments, and others like them, was to investigate that great mystery of how one cell generates a whole new organism. But the results also put a damper on fantasies about, say, cloning a political leader from cells of his nose. It appears that once a cell has become specialized enough to be part of a nose, it has long since lost its ability to become anything else.

Scott F. Gilbert, *Developmental Biology* (Sunderland, Mass.: Sinauer Associates, 1985), p. 314.

THE MYSTERIOUS NUMBER PI

If the distance across the diameter of a circle is 1 foot, the distance around that circle is approximately 3.141592 feet. That number is pi— that famous number from geometry, one of the most mysterious and remarkable numbers known. Pi relates the circumference of a circle to its diameter.

We have to say that pi is approximately 3.141592 because pi cannot be expressed exactly. The digits after the decimal point go on forever; that was proved in the 1700s.

Although mathematicians have proved that the digits of pi go on forever, no one knows what all those digits are. There are lots of formulas for calculating the digits of pi. Most of them involve some simple calculation, repeated over and over. Each repetition adds more digits. One recent calculation, done on a computer in Japan, gave the value of pi to over two hundred million digits.

Here's another strange property of the number pi: as far as anyone can tell, the digits are in completely random order. There's only one correct order, but there's no pattern to it. In the long run, every digit appears just as often as every other digit. And there's no way to guess what the two-hundred-million-and-first digit will be just by looking at the preceding two hundred million digits.

The randomness of the digits, and the fact that they go on forever, leads to another striking conclusion: any number pattern you can think of—your social security number, for instance—must be lurking somewhere among the known or unknown digits of pi.

In fact, if you assign a number to every word in the English language, then somewhere, probably trillions of trillions of digits beyond the decimal point, lies the sequence of numbers representing any book you care to name.

David Wells, *The Penguin Dictionary of Curious and Interesting Numbers* (New York: Penguin, 1986).

Petr Beckmann, *A History of Pi* (New York: St. Martin's, 1971).

"Following Pi down the Decimal Trail," *Science News* 133:215 (April 2, 1988).

Springs in the Cafeteria

You're in line at a cafeteria. You pass a stainless-steel cart that holds several stacks of plates. When you take a plate off the top of one stack, the other plates in the stack rise from below just far enough to present the next plate at the same height as the one you just took!

Underneath each stack of plates is a spring whose tension is adjusted to keep the top plate level with the top of the cart. Once the adjustment is made, the top plate will always be level with the top of the cart, no matter how many plates are in the stack.

Those cafeteria plate dispensers cleverly exploit a general property of springs: if you put twice as much force on a spring, it compresses, or stretches, twice as far. If you put twenty plates on the stack, their weight compresses the spring just twice as far as the weight of ten plates.

This property of springs was discovered about three hundred years ago by the English physicist Robert Hooke. Hooke wrote, "as the tension, so the force." Hooke noticed this principle at work in all kinds of devices that rely on springy materials: spring scales, bows, watches, and the vibrating parts of musical instruments.

In the twentieth century we've come to realize that this relationship, now called Hooke's Law, is a result of forces between atoms in solid material. If two atoms in a metal are pulled apart, they pull on each other. If you separate the atoms more, the force between them increases, in exact proportion. This is true as long as you don't pull the atoms too far.

Because of this pervasive relationship between atoms, just about all springs compress or stretch twice as far if you load them with twice the force. And because of that, cafeteria plate dispensers keep the top plate level with the top of the cart, no matter how many plates are in the stack.

Quotation from Hooke's treatise in the anthology *Moments of Discovery,* vol. 1, ed. George Schwartz and Philip W. Bishop (New York: Basic Books, 1958).

Richard P. Feynman, *The Feynman Lectures on Physics* (Reading, Mass.: Addison-Wesley, 1964), vol. I, chap. 12.

Firefly Signals

Firefly flashes are mating signals. Male fireflies cruise the evening air, flashing their lanterns in a pattern characteristic of their species, looking for females of their own kind.

The males of some species flash with a slow glow lasting several seconds; others flicker more than forty times per second. There are more than a hundred firefly species, each with a particular male flash pattern.

The female firefly emerges from her burrow after sunset and waits on the ground. When she detects the flash of a male of her species, she flashes back in a characteristic female pattern. Exchanging flashes, male and female find each other and mate. Then the female returns to her burrow, and the male returns to the air.

Sometimes, however, the male does not return to the air, because he is eaten by the female. Female fireflies of some species eat male fireflies of other species. They attract these males by detecting their flash patterns and sending a return flash pattern that imitates a female of the appropriate species. Some of these duplicitous insects have a repertoire of five or more different flash patterns.

So, every time a male firefly sees a responsive female flash in the grass, he faces a life-and-death dilemma: hesitate, and risk losing out to another male in the race to mate, or rush in and risk being eaten alive.

J. E. Lloyd, "Mimicry in the Sexual Signals of Fireflies," *Scientific American*, July 1981.

Mirages

On a hot, sunny day, the road in the distance seems to be covered with water. It reflects the blue of the sky. But when you get there, the mirage is gone.

You realize, of course, that there was never any water there in the first place. But the apparition was convincing; what you saw looked exactly like light reflected from a puddle. And as far as the reflection of light is concerned, there might as well have been water on the road.

The pavement, baking in the sun, is hot. The hot pavement heats the air near the surface to a much hotter temperature than the air higher up. A few inches above the pavement there's a boundary between hot air and cooler air—a boundary that can reflect light just like a water surface.

Light can be reflected from any boundary between one transparent medium and another. Take, for example, the boundary between air and water. If you look toward the far side of a quiet lake, you see trees and mountains beautifully reflected from the top of the water surface.

Light can be reflected from the bottom of a water surface too. If you're underwater in a swimming pool, wearing goggles so things are in focus, look up and toward the far side of the pool. If the water is quiet, you will see a mirror-like reflection from the underside of the water surface.

Light can be reflected from either side of a boundary between one transparent medium and another. The reflection is best if light hits the boundary at a glancing angle.

Back to the road: the boundary between hot air near the pavement and cooler air a few inches up reflects light exactly as a water surface would. From a distance, viewing the road at a glancing angle, it's hard to tell whether you're seeing water, or just reflection from the top of a layer of hot air—a mirage.

Marcel Minnaert, *The Nature of Light and Colour in the Open Air* (New York: Dover, 1954).

Seeing Yellow

Why does a mix of red light and green light look yellow?

Imagine the colors of the rainbow. There's a continuous range of colors, from red through orange, yellow, green, and blue, to violet.

The cells in our retinas that give us color sensations come in only three types. Each type is sensitive to a fairly broad segment of that rainbow—a fairly broad range of colors—with peak sensitivity in one region. Each cell generates a signal in proportion to the total amount of light it detects. Our nerves and brain combine those signals to give us color sensations.

The perception of the color yellow calls upon two types of these cells. One type has its peak sensitivity in the green range, with less sensitivity to colors near that peak, and the other has its peak sensitivity farther toward the red range.

Suppose pure light from the yellow part of the rainbow enters our eye. There's no color-detecting cell with peak sensitivity to yellow. But yellow light does stimulate the green-sensitive cells, slightly, and the red-sensitive cells, slightly. Signals from those cells are combined by our nerves and brain to give the sensation of yellow.

Now, if a mix of red and green light falls on our retina, it's the same thing as yellow as far as our visual system is concerned. Again, red-sensitive and green-sensitive cells are both stimulated, and again their signals are combined. Again, we get a sensation of yellow.

This simple picture shows that the job of detecting what we usually call yellow light is shared by two types of color-detecting cells in our eyes. Whether those two types are stimulated by pure yellow light, or by a mix or red and green light, the signals that finally come out are the same.

Other things being equal, our eyes cannot distinguish yellow light from a mix of red light and green light.

R. Bergsten, "When Is Yellow *Yellow?*" *The Physics Teacher*, October 1986.

WHY COTTON WRINKLES

A cool, dry cotton fiber springs back after being bent. A warm, damp cotton fiber doesn't. Temperature and moisture make the difference.

First, the role of temperature.

Cotton fibers, like many other materials, spring back as long as they're cooler than a certain transition temperature—the temperature at which the fiber loses its springy quality. This temperature of transition from springy to non-springy is a characteristic of the material. For dry cotton, this point is about 120 degrees Fahrenheit. Dry cotton stays springy below 120 degrees; above 120 degrees, it holds whatever shape it's bent into.

Now the role of moisture.

Water, absorbed into cotton fibers, lowers this transition temperature. In other words, damp cotton loses its springiness at a lower temperature than dry cotton.

Suppose the weather is warm and you're wearing a cotton shirt or blouse. You sit in a chair, leaning back. Perspiration lowers the transition temperature of the fabric to about 70 degrees Fahrenheit. Your skin is warmer than that, so heat from your skin is sufficient to change the cotton into its non-springy state. Your weight then presses the cotton into a new, wrinkled shape. You stand up; the cotton cools off, dries out, and resumes its springy state—this time in a wrinkled shape.

The cotton holds its wrinkled shape till you iron it flat again, probably with a steam iron. Steam lowers the transition temperature of the cotton to a point much cooler than the temperature of the iron. The iron can then press the fabric into a new, flat shape.

F. L. Wiseman, *Chemistry in the Modern World: Concepts and Applications* (New York: McGraw-Hill, 1985).

"Glass Transition," in *McGraw-Hill Encyclopedia of Science and Technology* (New York: McGraw-Hill, 1987).

DNA CAUGHT IN THE ACT

DNA, deoxyribonucleic acid, is the famous molecule that carries genetic information. A DNA molecule takes the form of a thread of two strands twisted around each other in a double helix.

Every time a cell divides, the DNA inside copies itself. The double helix unwinds. Each old strand serves as the template for a new companion strand assembled from raw materials in its vicinity. The result is two identical DNA molecules where there used to be only one. That's how genetic information is passed on.

This all-important copying process was actually photographed by the English biologist John Cairns in 1962. The experimental subjects were intestinal bacteria.

The strategy was this: feed the bacteria radioactive material that they'll incorporate into their DNA. Leave the bacteria alone for a while, then remove their DNA, spread it on a glass plate, and coat the plate with photographic emulsion. Wherever the DNA has taken up some of the radioactive material, it'll leave a mark on the emulsion. Develop the emulsion, examine it under a microscope, and you should be able to see photographic images left by the radioactive DNA molecules.

The experiment was technically difficult, but it worked. Cairns's 1962 photographs showed what looked like loops of black thread—those were the DNA molecules. And in some places, those threads divided into two strands. DNA had been caught in the all-important act of copying itself. The process that conveys hereditary information from generation to generation had been made visible.

John Cairns, "The Bacterial Chromosome," *Scientific American*, January 1966 (reprinted in the *Scientific American* anthology *Genetics*, intro. by C. I. Davern [San Francisco: W. H. Freeman, 1981]).

Newton's rainbow

Isaac Newton, the seventeenth-century English physicist, darkened a room by covering all the windows. Then he cut a small hole in one window shade to let in a narrow beam of sunlight.

He put a triangular solid-glass prism in the sunbeam. The prism bent the sunbeam and projected it onto the opposite wall of the room, twenty-two feet away. Where the beam hit the wall, it was spread out into an oblong patch with all the colors of the rainbow—a spectrum.

Others had noticed that when sunlight goes into a glass prism, colors come out. But how? Why?

One popular explanation in Newton's time was that the prism created the colors by somehow changing the white light into other forms. Newton's interpretation was different. He guessed that those rainbow colors were all present in the original white light, but they were mixed. The prism sorted them out. Today we know Newton was right, but he had to do a lot of thinking and experimenting to convince himself.

For example, Newton figured that if the prism was really taking the white light apart, then it should be possible to put the colors back together and make white light once again. In one of his simpler experiments, Newton projected not one but three spectra on the wall and made them overlap partially, to mix the colors. The result was white light, reconstituted from rainbow colors.

So Isaac Newton, working three centuries ago, did far more than just observe the colors of the spectrum; he proposed a new explanation for the colors, and he tested his explanation with experiments. Newton not only took white light apart, he put it back together.

I. B. Cohen, "Isaac Newton," in *Dictionary of Scientific Biography* (New York: Charles Scribner's Sons, 1974).

R. S. Westfall, *Never at Rest: A Biography of Isaac Newton* (New York: Cambridge University Press, 1980).

The elastic ruler

Of all the home demonstrations we've described in this book, this might be the simplest. All you need is a ruler and a saucepan full of water.

Hold the ruler vertically and lower it into the water on the side of the pan farthest from you. Bring your eye down to a position just a little higher than water level and look at the ruler. The submerged part of the ruler looks shorter than the part above water. Notice that the effect becomes more pronounced as you lower your eye closer and closer to the level of the water surface.

Now try this: fix your eye just a little above water level and slowly raise the ruler out of the water. The ruler seems to stretch as it emerges from the water. And, of course, it works in reverse: dip the ruler back in and it seems to compress.

The physical principle at work here is that light is bent whenever it crosses at an angle from one transparent medium into another.

You can try to imagine following light rays as they leave the bottom of the ruler on their way to your eye. Light rays leaving the bottom of the ruler at a steep angle—say, upward at forty-five degrees—get bent toward the horizontal when they leave the water. The light rays then travel through the air at a shallow angle, closer to horizontal. Light rays reaching your eye at a shallow angle appear to come from just beneath the surface of the water.

C. J. Lynde, *Science Experiences with Ten-Cent Store Equipment* (Scranton, Pa.: International Textbook Co., 1951), p. 71.

UNCORKING A MYSTERY

The seventeenth-century English physicist Robert Hooke was curious about the remarkable properties of cork—its ability to float, its springy quality, its usefulness in sealing bottles. He investigated the structure of cork with a new scientific instrument he was very enthusiastic about: the microscope.

Hooke cut a thin slice of cork with a penknife, put it under his microscope, focused sunlight on it with a thick lens, and looked through the eyepiece. What he saw looked like a piece of honeycomb. The cork was full of small empty compartments separated by thin walls. He called the compartments "pores, or cells." He estimated that every cubic inch of cork had about twelve hundred million of these cells.

Hooke had discovered the small-scale structure of cork. And he concluded that the small-scale structure of cork explained its large-scale properties.

Cork floats, Hooke reasoned, because air is sealed in the cells. That air springs back after being compressed, and that's why cork is springy. And that springiness, combined with the fact that the cells are sealed off from each other, explains why a piece of cork is so well suited for sealing a bottle.

Hooke's observation not only explained the properties of cork, but gave a hint that all living tissue might be made of small building blocks.

Our understanding of what those building blocks are has changed since Hooke's time. Today we'd say that what Hooke observed were dead walls that had been created by living cells when the cork was still part of the tree. But we still use the word "cell," and our usage can be traced back to the microscopic observations of cork made over three hundred years ago by Robert Hooke.

R. Hooke, *Micrographia* (New York: Dover, 1961).

R. S. Westfall, "Robert Hooke," in *Dictionary of Scientific Biography* (New York: Charles Scribner's Sons, 1972).

Psychological accounting

Here are two hypothetical situations involving money. What would you decide to do in each situation?

First, imagine that you have decided to see a play to which admission is ten dollars per ticket. As you enter the theater, you discover that you have lost a ten-dollar bill. Assuming you have more money, would you still pay ten dollars for a ticket to the play?

When a psychologist posed this question to over 180 people, most of them—88 percent—said they would go ahead and buy the ten-dollar ticket even after discovering that they had lost a ten-dollar bill.

In the second situation, you've already bought a ten-dollar ticket to the play. But as you enter the theater, you discover that you have lost your ticket. The seat is not marked; you can't recover the lost ticket. Would you pay ten dollars for another ticket?

When this question was posed to a couple of hundred people, only 46 percent said they would pay ten dollars for a replacement ticket.

In both situations, you end up paying twenty dollars, rather than ten, to see the play. The difference in people's responses seems to be a matter of psychological accounting. In the first case, the lost money and the cost of a ticket seem to have been entered in separate accounts; losing ten dollars had no particular bearing on the decision to buy a ticket to the play. In the second situation, the lost object was the ticket itself. The cost of the new ticket and the cost of the old ticket were entered in the same psychological account; the total cost of twenty dollars then seemed excessive.

This experiment gives evidence that what we decide to do depends not only on the information we have, but also on how we organize that information. Viewing the same question two different ways, we may come up with two different answers.

Amos Tversky and Daniel Kahnemann, "The Framing of Decisions and the Psychology of Choice," *Science* 211:453–458 (January 30, 1981).

COLOR IS IN THE EYE OF THE BEHOLDER

Black and white, plus black and red, make full color, because color *is* in the eye of the beholder. That was demonstrated in the 1950s by the engineer Edwin Land, also famous as the founder of the Polaroid Corporation.

Land's procedure was this: make two photographs of the same colorful scene—say, a vase of flowers—on black-and-white film. But make the photographs slightly different by taking one picture through red glass and the other through green glass. Load these two black-and-white photographs into two slide projectors pointed at the same screen. Superimpose the projected images to make a single black-and-white image.

Now put a piece of red glass in front of the projector containing the picture originally taken through red glass. You might think this would just turn the image on the screen pink. But, astonishingly, the image on the screen takes on a nearly full range of colors, including blues and greens—even though neither blue nor green light is hitting the screen. A black-and-red image plus a black-and-white image makes a full-color image!

This experiment, and others like it, showed that how we experience light depends on context. Our visual system assigns colors to objects in a picture on the basis of differences in the light coming from different parts of the scene.

Edwin Land demonstrated in the 1950s that color is not really a property of light itself. Instead, color is in the eye of the beholder; it's a sensation created by our visual system. Our eye makes a world of color from whatever light is available.

E. H. Land, "Experiments in Color Vision," *Scientific American*, May 1959.

WINE AND LIFE

Yeast from crushed grapes turns grape juice into wine by the process of fermentation. Sugar in the juice is changed into carbon dioxide gas and ethyl alcohol.

How does yeast do that?

A hundred years ago, a lot of scientists thought yeast caused the transformation of grape juice into wine through some "vital force" present only in living things like yeast cells. According to this "vitalistic" view, fermentation is inextricably linked with life. Other scientists had a different view, namely, that all the processes of living organisms could, in principle, be explained by the same laws of chemistry and physics that describe non-living things.

A crucial experiment was carried out in 1896 by the German chemist Eduard Buchner. He mixed yeast with sand and diatomaceous earth, ground the mixture in a mortar and pestle, put the mixture in a hydraulic press, crushed it under tremendous pressure, and squeezed out a fluid that had been part of the yeast cells.

When Buchner added sugar to this fluid, the sugar changed to alcohol, and bubbles of gas came out. Fermentation happened without yeast cells. Buchner had demonstrated that fermentation was caused not by some vital force in the yeast, but by a chemical, an enzyme, that can work by itself in a test tube.

Eduard Buchner was credited with opening up a new field of research: biochemistry. Buchner specifically rejected the vitalistic viewpoint—the view that the chemistry of life was unknowable. His demonstration of fermentation without living cells also earned him the 1907 Nobel Prize for Chemistry. And since Buchner's time, biological research has been based solidly on the notion that the same laws of physics and chemistry apply to living things and non-living things.

H. Schriefers, "Eduard Buchner," in *Dictionary of Scientific Biography* (New York: Charles Scribner's Sons, 1970).

E. Buchner, Nobel Lecture, 1907, with presentation speech by K. Mörner, in *Nobel Lectures: Chemistry, 1901–1921* (New York: Elsevier, 1966).

THE SWEET SPOT ON A BASEBALL BAT

There's a right way and a wrong way to hit a ball with a bat. If you hit the ball with the wrong part of the bat, your hands get stung and the bat may break. If you hit the ball on the "sweet spot" of the bat, you get the feeling of a solid connection with the ball—no stinging, no vibration, no broken bat.

Every bat has a sweet spot. The location along the length of the bat varies, depending on the shape of the bat and on how you hold it. You can find the sweet spot by gripping the bat handle with one hand in the same place as when you're swinging. Take a hammer in the other hand and gently tap the bat at various places along its length. At some point you will feel almost no vibration when you tap. That's the sweet spot. That's the best place for the ball to meet the bat.

If you don't have a baseball bat handy, you can find the sweet spot on a pencil. Hold the pencil loosely between two fingers and tap it sharply against the edge of a hard table. Try tapping various spots along the length of the pencil, from the far end to very close to your fingers. You'll find one point where the tapping does not sting your fingers. You'll probably also notice a difference in the sound as you get closer to the sweet spot.

Any swinging object has a sweet spot, or "center of percussion," as engineers call it. A good hammer or axe is designed so that if you grip the handle at the proper place, the sweet spot is right where the tool strikes the nail or the wood. As a result, you get the optimum in power and control.

P. Kirkpatrick, "Batting the Ball," *American Journal of Physics* 31:606 (1963).

U. Ingard and W. L. Kraushaar, *Introduction to Mechanics, Matter, and Waves* (Reading, Mass.: Addison-Wesley, 1960), p. 343.

O. Eschbach, *Handbook of Engineering Fundamentals*, 2nd ed. (New York: Wiley, 1969), pp. 4–41.

Make an image without a lens

Punch a pinhole in the center of a big piece of cardboard. That pinhole can project an upside-down image onto a piece of white paper.

The effect is easiest to see if you're indoors on a sunny day, in a room with a window. Stand on the side of the room farthest from the window. Hold the card with the pinhole vertically. Hold a piece of white paper in the shadow of the card, behind the pinhole. Light coming through the pinhole will make an upside-down image of the window on the white paper.

Hold the white paper closer to the pinhole, and you get a brighter but smaller image. Make the pinhole bigger by shoving a pencil point into it, and you get a brighter but fuzzier image.

A way to see how it works is to think of rays of light traveling in straight lines from each part of the window, through the pinhole, onto the white paper. Because the pinhole is small, each point on the white paper receives light from only a tiny part of the window. So light from the window is formed into an image on the white paper.

You can also see that light from the top of the window, after going through the hole, ends up at the bottom of the projected image. That's why the image is upside down.

Making an image with a pinhole is so simple, any child can do it. But the technique is also used on the cutting edge of modern science. Some objects in space, including supernovas and black holes, emit X-rays. Ordinary lenses and mirrors cannot focus X-rays, but a pinhole can. Many X-ray telescopes aboard astronomical satellites are elaborate pinhole cameras, based on the same principle you've just demonstrated with a piece of cardboard.

G. K. Skinner, "X-ray Imaging with Coded Masks," *Scientific American*, August 1988.

C. J. Lynde, *Science Experiences with Ten-Cent Store Equipment*, 2nd ed. (Scranton, Pa.: International Textbook Co., 1951).

CHIMES, FOR YOUR EARS ONLY

This kitchen demonstration calls for a fork and a piece of string about five feet long. Tie the middle of the string tightly around the fork at the narrowest part of the handle. Hold one end of the string in each hand and let the fork hang down in the middle. Press the ends of the string to your ears and swing the hanging fork against the edge of a table so it strikes once and bounces away. If the fork you're using is made of one piece of metal, with no joints or rivets, it will ring like a bell and send a surprisingly loud and rich chiming sound through the string to your ears.

If you try the same thing without pressing the string against your ears, the sound isn't nearly as rich. The reason for the loudness is that the taut string carries the fork's vibrations much better than air does. Take the string away from your ears, and all you hear is whatever comes to your eardrums through the air.

The reason for the richness of the sound is that you've suspended the fork in a way that allows it to vibrate freely, at many frequencies, making many musical tones at the same time. If you were holding the fork in your hand, the flesh on your fingers would damp out a lot of those vibrations.

Try tying not just one but two forks, or a fork and a spoon, into the middle of the string. Allow the cutlery to hang freely, press the ends of the string to your ears, and let the cutlery strike the edge of a table. From plain, ordinary flatware you can hear a sound almost like church bells.

C. J. Lynde, *Science Experiences with Ten-Cent Store Equipment,* 2nd ed. (Scranton, Pa.: International Textbook Co., 1951).

Martin Gardner, *Entertaining Science Experiments with Everyday Objects* (New York: Dover, 1981).

IT'S NOT JUST THE HEAT; IT'S THE HUMIDITY

Our bodies generate heat. The more active we are, the more heat we generate.

To keep our inner body temperature at the level where it belongs, we must dissipate that heat. If we don't, our body temperature begins to rise. We face heat stroke at a body temperature of about 106 degrees Fahrenheit—and death at a body temperature of about 110.

Heat always flows from warmer material into cooler material, never the other way. When the surrounding air is cooler than our body temperature, our skin is cooler than the inner body. That makes heat flow out, as it should.

If the air temperature gets close to our body temperature, something more is needed to keep the skin cooler than the inner body. We sweat.

When we sweat, energy represented by heat in our skin drives water molecules apart and sends them off into the air. In other words, evaporation cools our skin; the all-important temperature difference between inside and outside is restored, and heat continues to flow outward, even if the air temperature is close to our body temperature.

We're in trouble if the air is not only very hot but also humid. If the air is nearly saturated with water vapor, evaporation effectively stops, and sweating no longer cools our skin. Heat no longer flows out of our bodies, so our body temperature rises, especially if we continue to generate a lot of internal heat through hard work.

Heavy exertion, high temperature, and high relative humidity are a dangerous combination that no one can endure for long.

R. C. Plumb, "Knowing Some Thermodynamics Can Save a Life," *J. Chem. Ed.* 49 (2):112–113 (February 1972).

Fluorescent Lamps

In an incandescent bulb, electric current flows through a tungsten filament. In a fluorescent tube, current flows through rarefied gas.

Mixed in with that gas is a small amount of mercury vapor. Mercury atoms in the vapor pick up extra energy from the electric current. When those mercury atoms give up their extra energy, they give it up as ultraviolet light. That's a kind of light not visible to the eye, so the next step is to convert that ultraviolet to visible light.

Here's where the process of fluorescence comes in. Fluorescent materials convert one form of light into another. The inside of a fluorescent tube is coated with a fluorescent phosphor—a powder that absorbs energy in the form of ultraviolet light and re-emits that energy as visible light.

So in a fluorescent lamp, electrical energy is converted to ultraviolet light by mercury atoms; then the ultraviolet light is converted to visible light by the fluorescent phosphor on the inside of the tube.

Fluorescent tubes run cool. As lighting devices go, a fluorescent tube is very efficient at converting electrical energy into lots of light and very little heat. But before the electric current gets into the tube, it must go through a so-called ballast, a device that smooths out fluctuations in the current. The ballast is by far the most inefficient part of a fluorescent lamp.

"Fluorescent Lamp," in *McGraw-Hill Encyclopedia of Science and Technology* (New York: McGraw-Hill, 1987).

H. R. Crane, "Fluorescent Lights: A Few Basic Facts," *The Physics Teacher,* November 1985, p. 502.

"News Updates (Reagan Signs Ballast-Efficiency Standards)," *Science News* 134:30 (July 9, 1988).

J. Raloff, "Energy Efficiency: Less Means More—Fueling a Sustainable Future," *Science News* 133:296–298 (May 7, 1988).

THE TRUTH ABOUT BLEACH

Bleach doesn't work by taking anything out of fabric; it works by changing what's already there.

Take a grape juice stain, for example. Grape juice looks purple because of molecules in the juice that reflect purple light and absorb non-purple light. Those molecules absorb non-purple light because their atoms are connected by electrical bonds that store just the amounts of energy carried by non-purple light. What bleach does is to rearrange those bonds, those connections between atoms, so that they no longer absorb light as they did before. The formerly purple molecules are made colorless.

One important idea here is the relation between energy and color of light. The color sensation we get from light depends largely on how much energy the light carries. Different energies appear as different colors.

Another important idea is that molecules are made of atoms connected by electrical bonds. To say that a molecule absorbs light means that the molecule stores the energy brought to it by the light. The electrical bonds are the storage batteries. Different bonds hold different amounts of energy, so different bonds absorb light of different colors. Change the bonds between atoms and you change the molecule's color.

What bleach does is to rearrange the bonds between atoms so the formerly colored molecule doesn't absorb energy from light as it did before. The stuff that caused the stain is still in the fabric—bleach just makes it invisible.

F. L. Wiseman, *Chemistry in the Modern World* (New York: McGraw-Hill, 1985).

"Bleaching," in *McGraw-Hill Encyclopedia of Science and Technology* (New York: McGraw-Hill, 1987).

CARBON DATING

If an object contains plant or animal material of any kind—cotton, wool, or leather, for example—the technique of carbon dating may reveal the object's age.

Occasionally, subatomic particles from space hit nitrogen atoms in the Earth's atmosphere. The impact changes the nitrogen atoms into so-called carbon-14 atoms. A definite, small percentage of all the carbon atoms on our planet are carbon-14. They're pretty much like ordinary carbon atoms, but they're unstable: sooner or later, each carbon-14 atom changes back into nitrogen. It takes about six thousand years for half the carbon-14 atoms in any object to change back to nitrogen.

Plants and animals take in carbon, including carbon-14, from their environment while they live. Plants get carbon from carbon dioxide in the air; animals eat the plants. But when a plant or an animal dies, it stops taking in new carbon. In the dead body, the amount of ordinary carbon stays the same, while the carbon-14 gradually disappears—it keeps changing into nitrogen.

So, to determine the age of something by the carbon-14 method, you measure the total amount of carbon in the object, then measure how much of that is still in the form of carbon-14. The less carbon-14, the longer it has been since the death of whatever plant or animal the object was made from.

Carbon-14 dating works best for objects less than about fifty thousand years old. There are other dating methods that work for inanimate objects millions or billions of years old, using other unstable atoms.

"Radiocarbon Dating," in *McGraw-Hill Encyclopedia of Science and Technology* (New York: McGraw-Hill, 1987).

M. W. Rowe, "Radioactive Dating: A Method for Geochronology," *Journal of Chemical Education* 62:580–584 (July 1985).

R. F. Flint and B. J. Skinner, *Physical Geology* (New York: Wiley, 1974).

Soap Bubbles and Butterfly Wings

We can think of light as waves traveling through space, like ripples across a pond. Light can have different wavelengths—different distances between the crest of one wave and the crest of the next. Different wavelengths of light give us the sensation of different colors.

Those are some abstract but basic ideas about light. Here's how to see those ideas in action.

Make a solution of soapy water in a shallow bowl. Take a coffee cup and lower it, mouth down, into the soapy water, then lift it out gently. You want a thin soap-bubble film across the mouth of the cup.

Now turn the cup sideways so the soap-bubble film is vertical. Hold the cup so you can see the reflection of a bright light—the sky, for instance—in the surface of the bubble. After a few seconds you'll see bands of color appear near the top of the bubble, slowly moving downward.

The soapy water has no color of its own. The thickness of the bubble makes the colors. Wherever the bubble is thick enough to accommodate a whole number of waves of, say, red light, red light will be reflected to your eye, because red-light waves bouncing off the back side of the bubble will be exactly in step with red-light waves bouncing off the front of the bubble. The same goes for other colors. As the bubble's thickness changes, so do the colors.

This is the same process that makes colors in oil floating on water. And most of the blue colors in bird feathers and butterfly wings also come about in the same way—not through colored dyes, but through reflection of light from both sides of a thin film.

97

C. J. Lynde, *Science Experiences with Ten-Cent Store Equipment*, 2nd ed. (Scranton, Pa.: International Textbook Co., 1951).

Richard P. Feynman, *QED: The Strange Theory of Light and Matter* (Princeton: Princeton University Press, 1985).

TRACING THE ROOTS OF ENERGY

In 1841, the German physician Julius Mayer went to the East Indies as the doctor on a Dutch merchant ship. While treating European sailors in Java, he was struck by the color of their blood. Mayer was treating the sailors by bleeding, as physicians of the nineteenth century often did. He noticed that blood from the veins of these sailors had a much brighter red color than usual—almost as bright as blood from arteries.

When Mayer saw this, he boldly guessed that the sailors were using less blood oxygen than usual because they were in a hot, tropical climate new to them. In those days, physicians were beginning to realize that body heat resulted from something like slow burning of food and oxygen. Mayer guessed that since the sailors were getting so much heat from outside their bodies, they didn't need to make as much heat inside, so they used less of the oxygen in their blood.

Mayer's observations of the sailors' blood led him to guess that energy—he called it "force"—is never created or destroyed; it just changes from one form into another. The way Julius Mayer said this was vague and hard to evaluate in the 1840s, but his hunch contributed to an idea—the conservation of energy—that is now basic to all science.

Today, for example, we say that nuclear energy from the sun is sent to Earth in the form of light. Plants store it as chemical energy; animals, including us, eat the plants and change the stored chemical energy into energy of motion and into heat.

At each stage of the process, energy income and energy outgo must balance exactly, as in a bank account. Energy is a kind of common currency. Today we interpret every event in nature as the transformation of energy from one form to another.

R. S. Turner, "Julius Robert Mayer," in *Dictionary of Scientific Biography* (New York: Charles Scribner's Sons, 1974).

WHY FAN BLADES STAY DIRTY

Turn off your window fan for a moment, and you see fine dust on the fan blades. It seems strange—when the fan is running, there's air blowing over the blades hour after hour. Yet that wind doesn't carry the dust away.

When you look at dirt on fan blades, you're looking at a manifestation of one of the most complex processes in nature: air flowing over a solid surface. Figuring out the details requires the fastest supercomputers in existence. But the overall picture is something like this:

If you could get small enough to sit on a spinning fan blade, you'd feel a strong breeze—it would be like riding in a convertible with the top down. But here's the strange and mysterious part: as you got still smaller and closer to the surface, you'd feel less and less breeze. Within a fraction of a millimeter from the fan blade surface, you'd feel no breeze at all!

Because of friction between air and fan blade, there's a very thin layer of air next to the surface that doesn't move over the blade. Any dust particles small enough to stay within that quiet surface layer never feel the breeze, so they stay put.

This fact has other practical implications. Blowing on a phonograph record won't get rid of the very smallest dust particles, especially the ones at the bottom of the grooves. And blowing on a camera lens will never get it really clean. In both cases you have to use a brush that touches the surface to get the fine dust off.

In his autobiography *Slide Rule* (Portsmouth, N.H.: Heinemann, 1954), Nevil Shute Norway describes calm air near the surface of a 750–foot-long airship built in 1930: "When the ship was cruising at about sixty miles an hour, as soon as you got to the top, or horizontal, part of the hull you were in calm air crawling on your hands and knees; if you knelt up you felt a breeze on your head and shoulders. If you stood up the wind was strong. It was pleasant up there sitting by the fins on a fine sunny day" (p. 101).

Richard P. Feynman, *The Feynman Lectures on Physics* (Reading, Mass.: Addison-Wesley, 1964), vol. II, chap. 41.

A. Khurana, "Numerical Simulations Reveal Fluid Flows near Solid Boundaries, *Physics Today*, May 1988.

C. Donald Ahrens, *Meteorology Today,* 2nd ed. (St. Paul: West, 1985) (discussion of saltation, p. 260).

I. Peterson, "Reaching for the Supercomputing Moon," *Science News* 133:172 (March 12, 1988).

S. Weisburd, "Record Speedups for Parallel Processing," *Science News* 133:180 (March 19, 1988).

I. Peterson, "Friction Features," *Science News* 133:283 (April 30, 1988).

Brown Apples and Brown Tea

When you bite or cut into an apple, you crush some of the cells the apple is made of. Substances that normally occupy separate compartments within cells now have the chance to combine. Chemical reactions happen that would not happen in an uninjured apple, and one of those reactions leads to browning.

In particular, a family of substances called phenols from one compartment of each apple cell come into contact with oxygen from the air. When certain enzymes, the so-called phenolases, reach that mixture, a whole chain of chemical reactions takes place, and one of the end products is a brown-colored pigment. That pigment doesn't change the apple's taste or nutritional value, but it does hurt its visual appeal.

To prevent cut apples from browning, you must either inactivate the browning enzymes with heat or with acid such as lemon juice, or you must cut off the oxygen supply by immersing the fruit in water or by coating it with salad dressing. Vitamin C also prevents browning by starting a chemical reaction that uses up oxygen.

Browning also happens with bananas, pears, avocados, peaches, and raw potatoes, among other fruits and vegetables. The process apparently benefits the fruit, because it produces not only the brown pigment but also a substance that attacks fungi that might otherwise become established in the injured fruit tissue.

This "enzymic browning" has a practical application. In the processing of tea, the green leaves are crushed and exposed to the air for a few hours. Crushing of cells, mixing of substances, reaction with oxygen, and production of brown pigment all happen basically as they do with an apple. We don't like brown fruit, but we do like brown tea.

Harold McGee, *On Food and Cooking: The Science and Lore of the Kitchen* (New York: Macmillan, 1984).

T. P. Coultate, *Food—The Chemistry of Its Components* (London: Royal Society of Chemistry, 1984).

Owen R. Fennema, ed., *Food Chemistry*, 2nd ed. (New York: M. Dekker, 1985).

Optics and Glue

Fill a clean, dry glass with lukewarm water—lukewarm so that you don't get condensation of water from the air on the outside of the glass. Make sure your hand is completely dry, then pick up the glass and look down into the water, at the inside walls of the glass. You'll see a mirror-like reflection of the bottom. What you won't see is your hand holding the glass. You might see fingerprints here and there, but not much more. Wherever there's air between the glass and your skin, light is reflected back into the glass without ever reaching your skin.

Now dip your hand in water and pick the glass up with wet fingers. Look at the inside of the glass again and you'll see your fingers. When your hand is wet, water fills the tiny gaps between skin and glass, so light from i..side the glass can travel all the way out to your skin and back, allowing you to see your skin.

This not only demonstrates something about optics, it also shows why solid objects don't usually stick together: dry surfaces actually touch in only a few places.

Here's where glue comes in. It fills the gaps between objects and keeps them filled. Then the objects are held together by attraction between molecules, which is very strong if a lot of molecules are involved.

Water fills gaps in some materials, but it's not good glue because it runs out too easily. To glue your hand to the glass, you'd need either a stiff substance like the adhesive on duct tape, or something that hardens, like epoxy glue.

So, solid objects that seem to be in contact are actually separated by a lot of air. The job of glue is to fill in the gaps and keep them filled.

N. A. de Bruyne, "The Action of Adhesives," *Scientific American*, April 1962, and "How Glue Sticks," *Nature* 180:262–266 (August 10, 1957).

"Adhesive," in *McGraw-Hill Encyclopedia of Science and Technology* (New York: McGraw-Hill, 1987).

LIGHT TAKES TIME

In September 1676, the Danish astronomer Olaus Roemer went before the French Academy of Sciences in Paris and predicted that the eclipse of Jupiter's moon Io, which was expected the following November 9th at 5:25 A.M., would occur ten minutes late. His explanation for the delay was that earlier calculations of Io's motion had failed to consider the time needed for light to cross Earth's orbit. Roemer's prediction, and his explanation, turned out to be right.

Io is one of Jupiter's four largest moons. It's easy to see in a small telescope as it goes around Jupiter once every forty-two and a half hours. Eclipses of Io happen when it disappears into the shadow of Jupiter during part of each orbit.

The astronomers studying Io were working on ways to improve navigation methods. To steer by the stars, you have to know the time. They hoped to produce an almanac with the correct times of future eclipses of Io, so that navigators could check the time at sea by watching Jupiter and Io through a spyglass.

Roemer collected timings of Io's eclipses and saw an irregularity in the numbers: the interval between eclipses varied, depending on whether Earth was moving closer to Jupiter or farther away. Roemer concluded that light from Jupiter did not travel instantaneously as was generally believed, but took some time to reach Earth.

From those eclipse timings, Roemer calculated that light takes about twenty-two minutes to go all the way across Earth's orbit. He wasn't too far off: it actually takes about seventeen minutes, traveling at 186,000 miles per second.

Olaus Roemer is now remembered for having made, in 1676, the first careful measurement of a fundamental quantity of the universe: the speed of light.

Heinz Tobin, "Ferdinand Roemer," in *Dictionary of Scientific Biography* (New York: Charles Scribner's Sons, 1975).

I. B. Cohen, "Roemer and the First Determination of the Velocity of Light (1676)," *Isis* 31:327–379 (1940).

Read Fine Print through a Pinhole

Hold a page with fine print on it about an inch from your eye. Probably you won't be able to get the print in focus well enough to read it. The lens in your eyeball can't make a sharp image of an object that close.

Now try this: punch a small hole in an index card with a straight pin. Put down the pin and hold that pinhole very close to one eye—close enough so that the card touches your eyelashes. If you wear glasses, take them off.

Look through the pinhole at fine print an inch away. The letters will appear amazingly sharp. Take the pinhole away, and the view becomes blurry again.

The pinhole, being as small as it is, allows each point on your retina to get light only from one small part of the object you're looking at. As a result, all the light rays coming through the pinhole make an image of the fine print on your retina. That's how the pinhole makes such a big difference in what you see.

The image even looks magnified, because things look bigger when they're closer, and you're not accustomed to looking at things at such close range. This pseudo-magnification effect is more striking if you bring the print very close to the pinhole—a quarter of an inch or so. Keep the pinhole very close to your eye. Move the letters past the pinhole, and you'll see the middle of each letter bulge as it goes by.

So a pinhole functions something like a lens, but it works over a wider range of distances. Look through a pinhole and you can read fine print only an inch away.

C. J. Lynde, *Science Experiences with Ten-Cent Store Equipment,* 2nd ed. (Scranton, Pa.: International Textbook Co., 1951).

Get Your Bearings with Two Thumbtacks

Here's a way to learn a lot about your surroundings from one of those U.S. Geological Survey quadrant maps. They cover just a few square miles each, in tremendous detail. You can get one for your area from the

U.S. Government Printing Office or, possibly, from a local sporting-goods store.

Lay the map out flat on a table near a window. Find your location. Shove a thumbtack through that point on the map into the table below. Now the map is free to rotate around the thumbtack, but the point corresponding to your location stays fixed.

Next, look out the window and find some prominent landmark, preferably a few miles away—a hilltop, for example, or a fire tower. Find the symbol for that landmark on your map. Rotate the map around the thumbtack until a straight line drawn from the thumbtack through the map symbol points to the landmark. Stick one more thumbtack through that map symbol into the table. You've now oriented the map using two known points: your location, and a landmark you can see from your location.

Now that the direction to one landmark is correct, the directions to every other landmark will also be correct. A line of sight from the central thumbtack to, say, a hill whose name you don't know will pass through the symbol for that hill on your map. Look at the map, and you may find a name for the hill. Of course, if there's more than one hill along that line of sight, you may have to estimate the distance to the one you're looking at.

But you can get your bearings with two thumbtacks. Tack down your location first, then orient the map by referring to a landmark you can see. This works best with maps showing an area so small that the curvature of the Earth is not noticeable.

P. J. Davis and W. G. Chinn, *3.1416 and All That* (Boston: Birkhauser, 1985), p. 160.

A GLOBE AS A SUNDIAL

It's very easy to set up your globe as a miniature model of the real Earth, with the same orientation in space. Simply place your globe in a sunny spot, then tilt and turn it until your hometown is on top—so that a penny, laid flat on your hometown, doesn't slide off. Now turn the globe till the north side faces north. That's all there is to it. Leave the globe alone and let Earth's motion do the rest.

What you've done is to orient the globe so that its axis is parallel to the axis of Earth, and so that an imaginary line from your hometown on the globe, through the center of the globe, passes through your hometown on Earth and the center of Earth.

Now ponder how sunlight strikes the globe. The dividing line between sunlight and shadow on the globe corresponds exactly to the dividing line between the day side and the night side of Earth, right now. You can see where in the world the sun is rising, and where it's setting, at the present moment.

Since your globe is attached to the turning Earth, it will now turn every twenty-four hours. At different times of day, different times of year, different places in the world, the lighting on the little globe matches the lighting on the big Earth.

Anyone else in the world with a similarly oriented globe will see it lit exactly the same way as yours at exactly the same time. That person's globe, like yours, models the Earth we all share.

R. M. Sutton, "A Universal Sundial," in C. L. Strong, ed., *The Scientific American Book of Projects for the Amateur Scientist* (New York: Simon and Schuster, 1960), pp. 62–72.

An ink ring in a glass of water

For this experiment you need a tall glass of water (the taller the better), a bottle of ink, and an eyedropper.

After you fill the glass with water, put it on a steady table and leave it alone for a few minutes, to allow all the turbulence in the water to settle down. Now take some ink in the eyedropper and release one drop from a height of one inch above the center of the water. The ink will form an expanding ring, descending through the water like a smoke ring traveling through the air.

The ring keeps its shape because of the way water moves in it. You can get the idea by thinking of a rubber O-ring—maybe something like the rubber rings that are used as drive belts in upright vacuum cleaners. You can twist one of those rings inside out without changing its overall circular shape. What's happening with the water is something like that. The ink ring that you see is actually turning itself inside out as it travels to the bottom of the glass. The turning motion was started by the ink drop falling into the water.

The ring stays sharp and clear as it descends. That shows that water and ink in the ring don't mix very much with the rest of the water in the glass—at least not at first. There is some friction between the water in the ring and the surrounding water, and after traveling a long way, the ring will start to get fuzzy around the edges. But within the dimensions of a drinking glass, the ink ring will stay amazingly sharp and clear all the way to the bottom.

Richard M. Sutton, *Demonstration Experiments in Physics* (New York: McGraw-Hill, 1938), p. 103.
Richard P. Feynman, *The Feynman Lectures on Physics* (Reading, Mass.: Addison-Wesley, 1964), vol. II, chap. 40.

WHY AFTER-DINNER MINTS TASTE COOL

There are at least two reasons why chocolate-covered after-dinner mints make your mouth feel cool: the sudden melting of chocolate, and the strange cooling effect of menthol.

Most chocolate candy is about half cocoa butter. Cocoa butter is a natural component of cocoa beans and, like butter from milk, falls into the chemical category of fats and oils. But unlike butter from milk, butter from the cocoa bean has a remarkably uniform chemical composition; in other words, it contains only a few different kinds of fat molecules. That uniformity gives cocoa butter what chemists call a sharp melting point. As cocoa butter is heated, it changes suddenly from a brittle solid to a liquid as its temperature passes through about 93 degrees Fahrenheit.

When any substance melts, it absorbs heat from its surroundings. Anything that absorbs heat from your body when you touch it usually feels cool. When cocoa butter melts suddenly in your mouth, it absorbs a relatively large amount of heat in a short time, yet it does not get warmer right away. So you may notice that absorption of heat as a sensation of coolness.

The sudden cool sensation is especially noticeable because cocoa butter melts at about 93 degrees—very close to body temperature. That means that, just before it melts in your mouth, chocolate candy is likely to feel neither warm nor cold.

Now to the mint at the center of the candy. Mint flavor usually comes from oil of peppermint. The active ingredient in oil of peppermint is menthol, a fairly simple chemical with a remarkable effect: menthol tastes cool.

Apparently, menthol produces its cooling effect in the mouth not through the taste buds or even through the receptors for smell, but by fooling the cells that respond to temperature. In particular, menthol works on the receptors that detect cold, in your mouth or elsewhere. In small concentrations, menthol causes the sensors that detect cold to become active at a higher-than-usual temperature. So menthol makes things feel cooler than they really are. When you eat a mint containing menthol, the normal temperature of the inside of your mouth—body temperature—feels cool.

Menthol is also used in liqueurs, cigarettes, toothpaste, and shaving cream, among other products, to give a cool sensation to whatever part of the body the product touches.

P. W. Atkins, *Molecules* (New York: Scientific American, 1987).
Harold McGee, *On Food and Cooking: The Science and Lore of the Kitchen* (New York: Macmillan, 1984).
Owen R. Fennema, ed., *Food Chemistry*, 2nd ed. (New York: M. Dekker, 1985).

A LEAF FALLS

Where the stalk of a leaf meets the stem, there is a special layer of cells, the so-called separation layer. The separation cells are more closely spaced than cells in the surrounding tissue. In some plants, the separation layer is present even before the leaf matures; in other plants, those special cells develop later. Before the leaf falls, it dies, and whatever salvageable nutrients it contains are brought back into the tree.

When it's time for the leaf to fall, the cells of the separation layer produce enzymes that digest the walls between cells. Meanwhile, the separation cells closest to the plant get bigger, while those closest to the leaf don't. With the cell walls weakened and cells growing only on one side of the separation layer, the connection between leaf and stem soon breaks, and the leaf falls off.

Some of the separation layer is left behind on the stem. It forms a protective layer of cork that joins with the cork of the stem.

The whole process goes by the name of abscission, a word that comes from the same root as "scissors." Apparently abscission is controlled by a complex interaction among several plant hormones. With deciduous trees, it happens every year. Abscission can be triggered by cold, drought, decreasing day length, or polluted air, among other factors.

Left behind on the stem after abscission is a leaf scar with small raised dots inside it. Those dots are the so-called bundle scars, the sealed-off ends of the vessels that once carried fluids into and out of the leaf. Often, somewhere near the leaf scar, you can see the bud of next year's leaf.

"Abscission," in *McGraw-Hill Encyclopedia of Science and Technology* (New York: McGraw-Hill, 1987).

Frank B. Salisbury and Cleon W. Ross, *Plant Physiology*, 3rd ed. (Belmont, Calif.: Wadsworth, 1985).

A DOT, A LINE, A CREASE, A BEAUTIFUL CURVE

Take a piece of paper and a pencil. Draw a straight line somewhere on the paper—it doesn't matter where. Then draw one dot somewhere else on the paper—again, it doesn't matter where.

Now fold the paper over so that dot comes down somewhere on the line. Hold the dot on the line and crease the paper at the fold.

Open the paper up and re-fold it at a different angle. Put that dot somewhere else on the line, hold it there, and crease the paper at the fold.

Do this a few more times, always folding the paper at a different angle, but always putting that original dot somewhere on the line before you crease the paper.

Flatten the paper out, and you see that the creases form a curved boundary around the dot. The more creases, the smoother the curve. You can trace through the curve with a pencil.

Technically speaking, that curve is a parabola, one of the most famous curves of mathematics and physics. The path of a stream of water squirting out of a garden hose is close to a parabola; so is the path of a ball thrown through the air. Air resistance and the curvature of the Earth cause slight deviations from the parabolic ideal.

In space, where there's no air resistance, high-speed comets follow parabolic paths around the sun. The dot on your paper marks the position of the sun in the parabolic orbit. Make the dot and the line farther apart and you'll get a close-up of the portion of the comet's orbit nearest the sun.

Martin Gardner, "The Abstract Parabola Fits the Concrete World," *Scientific American*, August 1981.

DIRECTING A LIVING CELL

The cells of plants, animals, protozoans, and fungi contain a nucleus. To a large extent, the nucleus directs the cell. That was beautifully demonstrated in the 1930s by the German biologist Joachim Hämmerling, using single-celled plants known as *Acetabularia,* which grow in seawater.

Each plant is about one or two inches tall—small for a plant, huge for a single cell. One *Acetabularia* species looks something like a tiny palm tree. Another looks like a little parasol with an open canopy. Both types have a thin stalk, anchored to whatever the plant is growing on by a sort of foot. Within that "foot" is the cell nucleus.

Hämmerling cut the "foot" off one type of *Acetabularia* and grafted it onto a cut stalk of the other type. He was really transplanting the nucleus. After a few weeks, the stalk would grow a new cap.

What kind of cap? If Hämmerling put a palm-tree-type nucleus into a parasol-type stalk, that stalk would eventually grow a palm-tree cap. And a parasol nucleus in a palm-tree stalk would make a parasol cap.

Fifty years later, Hämmerling's experiments with *Acetabularia* plants are remembered for demonstrating the normal flow of information in cells—all kinds of cells. For the most part, the nucleus directs the cell. Information is transcribed from the famous DNA molecules in the nucleus into so-called RNA molecules, then translated into proteins which do the mechanical work of building.

J. Hämmerling, "Nucleo-cytoplasmic Relationships in the Development of Acetabularia," *International Review of Cytology* 2:475–498 (1953).

Scott F. Gilbert, *Developmental Biology* (Sunderland, Mass.: Sinauer Associates, 1985).

Dancing Pollen Grains

In the summer of 1827, the Scottish botanist Robert Brown looked through his microscope at pollen grains immersed in water. The grains were moving—they tumbled and jiggled back and forth as they drifted across the field of view. Brown wondered if the pollen grains were moving because they were alive.

Brown looked at pollen grains from dried plants that had been dead twenty years, at particles knocked loose from old wood and from coal, and at particles of soot from the air. They all moved when suspended in water. He then looked at particles that had never been part of anything living—dust knocked loose from rocks, including a rock supposedly taken from the Sphinx. The Brownian motion, as it's now called, was always there.

Clearly, this microscopic motion of tiny particles had nothing to do with life. But Brown couldn't explain it in 1827. After 1827, evidence accumulated indicating that all material bodies are made of atoms and molecules, and that temperature is a measure of the constant random motion of those atoms and molecules. The hotter the temperature, the faster the motion. The Brownian motion turned out to be caused by randomly moving water molecules knocking pollen grains back and forth on the microscope slide.

By 1905 the full significance of the Brownian motion was explained by a young physicist who showed that by watching how far a particle gets kicked by the Brownian motion in a certain amount of time, you can calculate how many molecules there are in a cubic centimeter of water. That young physicist was Albert Einstein.

Robert Brown, "A Brief Account of Microscopical Observations . . . on the Particles Contained in the Pollen of Plants . . . "; and A. Einstein, "The Elementary Theory of the Brownian Motion," both reprinted in Henry A. Boorse and Lloyd Motz, eds., *The World of the Atom* (New York: Basic Books, 1966).

Wet Weather Means Brighter Colors

There are several things you associate with dry summers, but one which you might not have noticed is that they are not as colorful. When wet weather returns, so do some of the colors we have been missing. Not only leaves, but rocks, soil, and road surfaces all have more intense color when they're wet than when they're dry.

Take a wet reddish-colored rock, for example. A wet rock is coated with a film of water. The water is smooth on the surface, and it fills all the tiny valleys and pits in the surface of the rock.

When light from the sky strikes this wet reddish-colored rock, the water bends the light downward, into those tiny valleys and pits. The light gets reflected back and forth from one side of each little valley to the other. At each bounce, more non-reddish-colored light is absorbed by the rock surface. The light that bounces becomes fainter and redder. A small amount of light bounces back up to the surface of the water film, and some of that light passes through the water surface into the air on its way to your eye. So when you look at a wet rock, you're seeing light that has bounced off the rock surface many times. A wet rock looks dark and strongly colored.

In the case of a dry rock, there's no water to bend light downward into the little valleys and pits. When you look at a dry rock, you're seeing light that has bounced off the rock surface only once, or just a few times. Relatively little light is absorbed by the rock surface, so the light that reaches your eye from a dry rock is whiter and more intense than light from a wet rock.

Because water fills the nooks and crannies in the surface, not only rocks but pavement, sand, soil, dead leaves, and bare wood look darker and more strongly colored when they're wet.

Marcel Minnaert, *The Nature of Light and Colour in the Open Air* (New York: Dover, 1954), p. 345.

THE SPINNING EARTH AND THE WEATHER

You and a friend sit on opposite sides of a big, flat turntable—a merry-go-round. With the turntable not moving, you toss a ball to your friend on the other side. If you toss accurately, the ball goes to the other person, who catches it.

Now the turntable begins to spin counterclockwise. Because of the turning, you begin moving to your right. You toss the ball to your friend on the other side once again. If you aim as you did before, you'll miss; the ball will turn right from your point of view. Your friend on the other side will have to lunge to his or her left to catch the ball. Because you're moving, the velocity of the turntable combines with the velocity of your throwing arm to send the ball off to the right. (If the turntable were spinning clockwise, the ball would turn left after you threw it.)

This effect was first analyzed in detail in 1835 by the French physicist Gaspard Gustave de Coriolis, who was making a theoretical study of the forces on moving parts of machines. Now we understand the tremendous importance of the Coriolis effect, as it has come to be called, in explaining how things move long distances over the Earth—air masses, for instance.

The Earth is, in effect, a giant turntable. As seen from the North Pole, the Earth spins counterclockwise every twenty-four hours. Because of that spinning, air flowing out from a northern-hemisphere high-pressure area turns right, just like the ball leaving your hand, and that causes clockwise wind circulation.

The Coriolis effect of the Earth's rotation is noticeable only with things traveling very long distances—things like winds and ocean currents. Contrary to what we sometimes hear, it's too weak to have a noticeable effect on water going down the drain in a sink.

Pierre Costabel, "Gaspard Gustave de Coriolis," in *Dictionary of Scientific Biography* (New York: Charles Scribner's Sons, 1971).

C. Donald Ahrens, *Meteorology Today*, 2nd ed. (St. Paul: West, 1985).

J. B. Marion, *Classical Dynamics of Particles and Systems,* 2nd ed. (Orlando: Academic Press, 1970).

INFECTION: A STRUGGLE BETWEEN TWO ORGANISMS

By 1882, the Russian biologist Elie Metchnikoff had collected many observations of how cells deal with foreign particles. Roundworms, for example, digested food with special cells that engulfed food particles. And white blood cells had been seen eating particles of the coloring matter in ink. Metchnikoff's observations led him to the hunch that animals, including humans, defend themselves against infection with special moving cells that devour invading objects.

One day in 1882, while his family was at the circus, Metchnikoff had an idea. As he described it, he guessed that "a splinter introduced into the body of the starfish larva . . . should soon be surrounded by mobile cells, as is to be observed in a man who runs a splinter into his finger."

Outside, in the garden, was a tree decorated for Christmas. Metchnikoff "fetched from it a few rose thorns and introduced them at once under the skin of the beautiful starfish larva, transparent as water." By the next morning, special cells—called "eating cells" in those days—had gathered around the splinter. The starfish larva was defending itself against the splinter.

After this 1882 experiment, Metchnikoff was convinced that infection was really "a struggle between two organisms."

Metchnikoff was a pioneer in the study of what is now called cell-mediated immunity, in which special cells gather around and attack intruders. There's another kind of immunity—antibody-mediated immunity—in which special molecules, the antibodies, circulate in the bloodstream in search of foreign molecules.

E. Metchnikoff, *Lectures on the Comparative Pathology of Inflammation* (1891; New York: Dover, 1968).

Quote taken from Olga Metchnikoff, *Life of Elie Metchnikoff* (1921; reprint, New York: Arno, 1972).

G. H. Brieger, "Elie Metchnikoff," in *Dictionary of Scientific Biography* (New York: Charles Scribner's Sons, 1974).

WHY DOESN'T A PREGNANT WOMAN REJECT HER FETUS?

The human body has ways of dealing with cells that don't belong to it. Sometimes the body rejects organ transplants and tissue grafts because they are made of foreign cells.

This leads to an interesting question. An unborn fetus is made of tissue foreign to the mother. The fetus is "foreign" because about half its genes come from the father, and its proteins are assembled according

113

to its unique genetic program. So why doesn't a mother's immune system reject her fetus?

The answer seems to be that part of the mother's immune response is suppressed while she's pregnant. Understanding of this has come about fairly recently, but evidence had been accumulating for a long time.

For example, decades ago in the rural southern United States, pregnant women were given quinine every day to try to protect them from malaria; and women who had had tuberculosis as young girls were considered completely cured if the tuberculosis did not return when they became pregnant. Behind these traditions was a recognition that resistance to these diseases is weak during pregnancy. Meanwhile, farmers and veterinarians had noticed that sheep had more worms when they were pregnant than when they were not.

More recently, experiments with mice have shown that pregnant mice are more susceptible to malaria than non-pregnant mice. And non-pregnant mice injected with hormones to imitate pregnancy are more susceptible to herpes than mice in a control group.

Why doesn't a pregnant woman—or any pregnant mammal—reject her fetus? Because her immune response is, so to speak, turned down as long as she's pregnant. The immune system must compromise between protecting the mother and protecting the fetus.

Eugene D. Weinberg, "Pregnancy-Associated Immune Suppression: Risks and Mechanisms," *Microbial Pathogenesis* 3:393–397 (1987).

THE FLOATING CORK TRICK

Cut a slice about a quarter of an inch thick from the end of a cork and drop it into a glass of water. Watch the cork for a few seconds and you'll see it drift over to the side of the glass. Challenge everyone at the table to make the cork float exactly in the center of the water. Let people push the cork around, turn it over, drop it in in some special way. No matter what they try, it will always drift to one side, for the uninitiated.

The trick is to leave the cork in the glass and add more water. Pour it in slowly from another glass. Keep pouring slowly and carefully until the water surface bulges above the rim of the original glass. You may be surprised at how much water will fit into that bulge without spilling.

The water bulges because molecules of water attract each other. The molecules at a water surface make a film under tension. That film of surface tension holds the water like a bag and keeps it from spilling. The same thing happens in a water drop.

Bring your eye down level with the rim of the glass and look at the profile of the water. You'll notice that it curves gently across the mouth of the glass, with the highest point of the bulge at the center. By this time you'll also notice that the buoyancy of the cork causes it to be pulled up the sloping water surface to that central high point.

Why didn't the cork float in the center before the glass was full? Pour some water from the glass and look carefully at the surface. Now you won't see the bulge in the center, but you can see that the water climbs up the sides a little all around the edge. This is because the water molecules are attracted to the glass. When the glass is not full, the water is higher at the edges than in the center, so the buoyancy of the cork still causes it to float to the highest point, but that is now at the edge rather than the center.

Martin Gardner, *Entertaining Science Experiments with Everyday Objects* (New York: Dover, 1981).

A THOUSANDTH OF A SECOND

A thousandth of a second is the shortest exposure time, the fastest shutter speed, on many cameras. That's especially helpful to sports photographers, who use fast shutter speeds to take pictures of high-speed motion. A runner who covers a hundred meters in about ten seconds goes only about a third of an inch in a thousandth of a second, so the runner's image will hardly be blurred in that short time.

In a thousandth of a second, a car going 55 miles per hour moves about an inch. A jet plane flying at 600 miles per hour goes about a foot. The strings in a piano that make the note middle C go through one-quarter of a complete back-and-forth vibration.

In the human body, a single molecule of the enzyme carbonic anhydrase makes about 100 molecules of carbonic acid from carbon dioxide and water in a thousandth of a second. That process helps regulate the amount of carbon dioxide in the bloodstream.

In a thousandth of a second, a mosquito flaps its wings once.

In a thousandth of a second, the electronic clock inside many personal computers makes about 15,000 ticks, each of which triggers another logical operation—that's why computers can do arithmetic so fast.

Here are some other things that happen in a thousandth of a second:

A beam of light or a radio wave travels about 186 miles—roughly the distance from Chicago to Indianapolis.

A spacecraft in low Earth orbit travels about 25 feet.

The Earth travels about 100 feet in its orbit around the sun; the planet Pluto travels about 17 feet in its orbit.

And our solar system gets about 46 feet closer to the star Vega and travels about 740 feet in its great orbit around the center of the Milky Way galaxy—in the brief interval of one one-thousandth of a second.

Based on a discussion in A. Bakst, *Mathematics: Its Magic and Mystery*, 3rd ed. (Princeton, N.J.: Van Nostrand, 1967).

The enzyme statistic comes from Helena Curtis, *Biology*, 4th ed. (New York: Worth, 1983), p. 168.

Touching and Being Touched

Our so-called sense of touch has at least two parts: passive touch, which tells us, for example, that something has landed on our arm; and active touch, in which we use the sensations of pressure on our skin to explore the world.

One of the first scientific studies of the difference between active and passive touch was done almost thirty years ago by the psychologist James Gibson. The study measured people's ability to distinguish shapes of six different cookie cutters—a triangle, a star, a crescent, and so on—without looking.

People's judgment of the shape was least accurate when the cookie cutter was steadily pressed into their outstretched palms by a mechanical lever—passive touching. People identified the shape correctly only about 29 percent of the time.

The subjects did better if the cookie cutter was attached to a mechanical gadget that slowly rotated it around a vertical axis while pressing it into their palms. About 72 percent of the judgments were correct.

The highest percentage of correct judgments, 95 percent, came when the cookie cutter was held still and the subjects were allowed to move their fingers to explore the shape—active touching.

These early experiments by James Gibson revealed something almost paradoxical. When we hold our hand still and allow something to touch it, the pressure on our hand is an exact image of the shape of the object; yet we usually can't tell what that shape is. But if we move our fingers over the object, we can form a very accurate idea of the shape, even though the pattern of pressure on our fingers has no resemblance to the shape. As Gibson wrote, "a clear unchanging perception arises when the flow of sense impressions changes most."

J. Gibson, "Observations on Active Touch," *Psychological Review* 69:477–491 (1962).
Philip G. Zimbardo, *Psychology and Life,* 12th ed. (Glenview, Ill.: Scott, Foresman, 1988).

ON A CLEAR DAY, HOW FAR CAN YOU SEE?

How far you can see depends on the condition of the atmosphere and on whether anything is blocking your view. But if we assume that the air is perfectly clear and the horizon is unobstructed, how far can you see?

Because the Earth is round, the higher you are, the farther you see. In case you're the calculating type, here's the formula: multiply the square root of your height, in feet, by a factor of one and a quarter; that gives the approximate number of miles to your horizon.

In case you're not the calculating type, here are some results:

If your eyes are five feet above ground level, your horizon is about two and three-quarters miles away.

If you're on the tenth floor of a tall building—about a hundred feet up—your horizon is about twelve miles away.

From fourteen hundred feet up—that's roughly the height of the Empire State Building—you can see forty-six miles if the air is perfectly clear.

From a jet at thirty thousand feet you can see about 210 miles; from a spacecraft a hundred miles up, about 890 miles.

That formula once again: multiply the square root of your height in feet by a factor of one and a quarter. That gives you the approximate number of miles to your horizon—on the Earth.

Elsewhere, the formula changes. On the moon, for instance, if your eyes are five feet off the ground you can see only about a mile and four-tenths. Since the moon is smaller than Earth, the surface curves more sharply and the horizon is closer.

If you could stand on the sun, you'd be standing on a surface more nearly flat. From five feet above the sun, with nothing blocking your view, you'd see thirty-one miles.

The flatter the surface, the farther you see. If the Earth were absolutely flat, then on a clear day you really could see forever.

A. Bakst, *Mathematics: Its Magic and Mystery,* 3rd ed. (Princeton, N.J.: Van Nostrand, 1967), chap. 29.

Edison and the Bulge-heads

In 1880, while working on his early carbon-filament electric light bulbs, Thomas Edison noticed something strange: after a bulb had burned for a while, a black deposit appeared on the inside. Edison figured that particles of carbon were being carried from the hot filament to the glass. He attacked the problem by sealing another wire into the bulb—a platinum wire connected to one terminal of the power supply, to attract the carbon particles.

When the improved light bulb was turned on, an electric current flowed through that extra platinum wire, even though the extra wire was not part of a complete circuit. Evidently electricity was flowing through empty space inside the bulb, from the regular filament to this extra wire! At that time no one would have imagined this was possible. A visiting English engineer named the phenomenon the Edison effect.

Not until the turn of the century did it become known that there were elementary particles of electricity (such as electrons) that could travel through empty space.

Thomas Edison apparently had little interest in either pure science or scientists. He patented the special bulbs and went on to other things. In a letter about the Edison effect, he wrote, "I have never had time to go into the aesthetic [i.e., theoretical] part of my work. But it has, I am told, a very important bearing on some laws now being formulated by the Bulge-headed fraternity."

In fact, electricity flowing through empty space can be controlled with much greater speed and precision than electricity flowing through a wire. When Edison made his contemptuous remark about "the bulge-headed fraternity," he was unaware of the implication of his discovery. His light bulbs with the extra wire inside were the first electronic devices; they would lead to the vacuum tubes that would make radio possible.

Matthew Josephson, *Edison* (New York: McGraw-Hill, 1959), pp. 274ff.

THE SHAPE OF SOUND

For this kitchen demonstration you need a candle, a piece of paper, and a big knife. Roll the paper around the candle, then put the candle on a cutting board and cut it in two with the knife. But cut it diagonally, like a green bean—in other words, cut straight down, but with the knife at an angle to the candle, not perpendicular to it.

Now unroll the paper and look at its cut edge. It's shaped in an undulating curve like the profile of ripples on a pond. That's one of the most important curves of mathematics and physics, the so-called sine curve.

Sine curves describe the nature of sound. The wavy edge of your paper could represent air pressure in a series of sound waves. Where the curve goes up, the air pressure is slightly higher than average; where the curve goes down, pressure is lower. When regular alternations of high and low pressure strike your eardrum, you hear a musical tone.

Look at the wavy edge of your paper and notice how far apart the waves are—the wavelength, as a physicist or engineer would call it. The distance between waves corresponds to the musical pitch of the sound. If the waves are closer together—in other words, if the wavelength is shorter—you hear a higher pitch. You can get a shorter wavelength by wrapping paper around a thinner candle and cutting it as you did before.

Look at the wavy edge and notice how far up and down the waves go.

That corresponds to the loudness of the sound wave. You can vary that by changing the angle of the knife when you cut through the candle. If the knife is more nearly parallel to the candle, you'll get a sine curve with higher highs and lower lows.

Hugo Steinhaus, *Mathematical Snapshots* (New York: Oxford University Press, 1950).

THE WRONG FORMULA FOR WATER

Most of us have heard that the chemical formula for water is H_2O. In other words, a molecule of water is made of two hydrogen atoms and one oxygen atom. But for a long time in the nineteenth century, chemists thought the formula was HO: one atom of hydrogen and one atom of oxygen in each water molecule.

That incorrect formula arose from the work of the nineteenth-century English chemist John Dalton, who published *A New System of Chemical Philosophy* in 1842. Dalton's reasoning went like this:

He was convinced that matter was made of atoms of different kinds: hydrogen was made of hydrogen atoms, oxygen was made of oxygen atoms, and so on. Dalton was also convinced that atoms could hook up in different combinations to make molecules, and that's why there are so many different chemical substances in the world.

In Dalton's time, chemists knew of only one substance that could be made from hydrogen and oxygen: water. Dalton figured, therefore, that a molecule of water was made of one atom of hydrogen and one atom of oxygen, because that was the simplest possible combination. The formula must be HO.

Dalton's reasoning was sound, but he was missing some important facts. He didn't know—no one knew in 1842—that water is not the only chemical that can form from hydrogen and oxygen. Dalton's mistake was in assuming that hydrogen and oxygen would combine in only one simple way.

John Dalton is one of the great figures in the history of science. His work in the early 1800s provided important evidence about the nature of atoms and how they combine. But because certain laboratory discoveries about hydrogen and oxygen had not yet been made in his time, Dalton came up with the wrong formula for water.

Aaron J. Ihde, *The Development of Modern Chemistry* (New York: Harper and Row, 1964).

J. Dalton, excerpt from *New System of Chemical Philosophy,* reprinted in Henry A. Boorse and Lloyd Motz, eds., *The World of the Atom* (New York: Basic Books, 1966).

DISCOVERING VIRUSES

Virus is Latin for "poisonous fluid." That's the guise under which viruses first became apparent a century ago.

In 1890 the Russian Department of Agriculture asked the botanist Dmitri Iosifovich Ivanovsky to investigate a so-called mosaic disease wiping out tobacco plants on the Crimean peninsula. Ivanovsky found that just the sap from an infected plant could cause mosaic disease in a previously healthy plant. Remarkably, even if all bacteria were filtered out, the sap could still make a healthy plant sick. Ivanovsky guessed that some living particle smaller than any known bacterium was causing the disease.

Meanwhile, in Holland, Willem Martinus Beijerinck was also studying tobacco mosaic disease in the 1890s. Beijerinck also found that filtered sap could infect a previously healthy tobacco plant. Sap from that second plant could then infect a third, and so on indefinitely. Beijerinck guessed that the virus, as he called the mysterious infectious agent, reproduced by using the reproductive machinery in plant cells.

Today, both Dmitri Ivanovsky and Martinus Beijerinck are credited with discovering viruses in the 1890s. But it took decades more to figure out how viruses really operate. Biologists now understand a virus to be a particle, much smaller and simpler than a cell, that attacks a preferred type of target cell and causes the chemical machinery inside that cell to make new viruses. Viruses reproduce only inside living cells. As Ivanovsky and Beijerinck guessed, viruses are unlike any other living organism.

V. Gutina, "Dmitri Iosifovich Ivanovsky," and S. S. Hughes, "Martinus Willem Beijerinck," in *Dictionary of Scientific Biography* (New York: Charles Scribner's Sons, 1973, 1978).

G. R. Taylor, *The Science of Life: A Picture History of Biology* (New York: McGraw-Hill, 1963).

Cells Get Old and Die

All living cells come from other cells. A cell makes new cells by dividing. But experiments first done in the 1960s showed that some cells won't keep dividing forever.

Some cells that make soft tissues of the human body, for instance, won't divide more than about fifty times. This was observed in experiments in which human tissue cells were cultivated in laboratory glassware. The cells somehow count divisions and stop dividing around the fiftieth division, even if they have nourishment and room to grow. After that, the cells die. Temperature and other factors influence the time interval between divisions, but not the final number of divisions.

Do we get old and die because our individual cells get old and die after fifty divisions? That's an intriguing thought, but the fifty-division limit can't be the whole explanation. For one thing, when old people die, many of their cells haven't yet reached the fifty-division limit. Also, that fifty-division limit was observed in a type of cell that divides especially frequently. Many of the changes associated with aging happen in types of cells that divide much less often. So the real significance of the limit on cell divisions is still unknown.

There are some cells that have a kind of immortality—they'll keep dividing indefinitely, no matter how crowded they are, as long as they get nourishment. They are cancer cells.

L. Hayflick, "Cell Senescence and Death," in *McGraw-Hill Encyclopedia of Science and Technology* (New York: McGraw-Hill, 1987).

"The Cell Biology of Human Aging," *Scientific American*, January 1980.

R. A. Weinberg, "Finding the Anti-oncogene," *Scientific American*, September 1988.

Sorting Out Musical Pitches

Sing into the strings of a piano with the damper pedal held down, and you hear reverberation of the notes you just sang. Pressing the pedal lifts the dampers from the strings, leaving the strings free to vibrate. The sound of your voice then causes the strings to vibrate—but not all equally. The strings that vibrate most energetically, and that keep vibrating after you stop singing, are the ones tuned to the pitches you sang.

Something a little like that happens in the human ear. Curled up in a fluid-filled capsule in the inner ear is a piece of tissue about an inch and a half long, the so-called basilar membrane, which is free to move in response to vibrations that come to it from outside. Just as different strings in the piano vibrate at different frequencies and make different musical pitches, different parts of the basilar membrane vibrate most energetically at different frequencies.

But there's a difference between the basilar membrane and a piano. The basilar membrane is one solid piece of elastic tissue, precisely shaped—thick at one end and thin at the other, like a tiny chisel. The basilar membrane does not have tuned strings; it has this special shape enabling it to sort out pitches.

The most energetic vibration is at the thick end for high frequencies, and at the thin end for low frequencies. Nerves connected to the membrane send signals to the brain indicating which part is vibrating most energetically.

If two or more musical notes sound together, the basilar membrane will vibrate strongly in two or more different places. That's why we can pick out individual notes in a musical chord. The basilar membrane of the inner ear, because of its special shape, sorts out vibrations by frequency.

The highest pitches humans can hear correspond to about twenty thousand vibrations per second. How do we do that? Not by counting vibrations.

A nerve, such as the nerve connecting the ear to the brain, cannot transmit twenty thousand impulses per second—in fact, a nerve may have trouble transmitting more than about five hundred or one thousand impulses per second.

Here is how our ear and brain deal with this inability of nerves to transmit such high frequencies. One area of the basilar membrane resonates at twenty thousand vibrations per second. Nerves connected to that area send the brain not twenty thousand nerve impulses per second, but a message that stands for twenty thousand vibrations per second. The brain then translates that message into a pitch sensation.

Incidentally, low notes—like a bass note at one hundred vibrations per second—are handled differently. For a low note, our ear does send one impulse to our brain for each and every vibration of the eardrum— one hundred impulses per second for that bass note.

J. G. Roederer, *Introduction to the Physics and Psychophysics of Music*, 2nd ed. (New York: Springer-Verlag, 1975).

THE ECHO OF A TRAIN

If you stand near a railroad track while a train goes by with its horn blowing, you hear the pitch of the horn drop as the train passes.

Imagine sound as waves of slightly compressed air, emitted from the horn, traveling through the air. The pitch you hear depends on how many sound waves reach your eardrum every second. More sound waves per second means higher pitch.

As the train approaches, the horn emits each new sound wave when it's a little closer to you than it was for the one before. So the sound waves are crowded together when they get to your ear; you hear a higher pitch than if the train were standing still.

As the train goes away, the horn emits each new sound wave a little farther from you than the one before; the sound waves are relatively spread out when they get to your ear, and you hear a lower pitch.

So, between the time of approaching and the time of going away, as the train passes, the horn's pitch drops: a familiar case of the so-called Doppler effect.

Try listening for this after the train passes: listen to the horn's echo from surfaces farther down the track. The echo has a higher pitch than the sound directly from the horn. The surface reflecting the sound to make the echo is getting crowded-together sound waves, because the train is approaching that distant surface. That reflecting surface isn't moving, so it doesn't change the pitch before returning the sound to you.

So when a train is going away from you, listen for the horn honking at one pitch, followed by an echo of that honk at a higher pitch.

MICROBURSTS AND AIRPLANE CRASHES

Turn on the kitchen faucet; watch the water hit the sink below and spread out in every direction. You're looking at a model of one of the principal causes of airplane crashes. Descending water from the faucet represents an intense, localized downdraft in the air—a so-called microburst—that comes out of some storm clouds.

Take an imaginary trip across the bottom of your sink while the water is running. As you approach the descending water column, you're going upstream, into the current. Then you get right under the faucet and feel a tremendous force pushing you downward. Finally, you emerge from the other side of the descending water column. The current is coming from behind you now; you're going downstream.

This analogy shows why it's dangerous to fly through a microburst in the atmosphere. First, the airplane approaching a microburst encounters a strong headwind that increases the speed of air over the wings, increasing the lift. Recall that airplane wings generate lift only if air flows over them at high speed. Then, suddenly, the plane hits a hundred-mile-an-hour downdraft that shoves it toward the ground. A few seconds later, the plane gets a strong tailwind that reduces the speed of air over the wings, undermining the lift and robbing the pilot of control of the aircraft. An airplane thousands of feet up may have time to recover from this; an airplane about to land may not.

Instruments are being installed on airplanes and at some airports to detect microbursts so that pilots and traffic controllers can better avoid them. But even with these instruments, airplanes will have to continue to avoid situations where microbursts are even likely.

Donald Ahrens, *Meteorology Today*, 3rd ed. (St. Paul: West, 1988).
"Airlines Told to Arm Planes against Wind Threat," *New York Times*, Friday, September 22, 1988, p. 13.

WHAT'S INSIDE AN ATOM?

If you were to wrap a steel marble in fluffy cotton, then shoot BBs at it, most of the BBs would go right through the cotton. But a few would hit that steel ball at the center and bounce back. Those ricocheting BBs would tell you that something small, heavy, and hard was inside that cotton. That's basically how the nucleus of the atom was discovered back in 1911.

Hans Geiger, Ernest Marsden, and Ernest Rutherford at the University of Manchester in England shot tiny particles at atoms and watched where the particles went. The tiny particles came from radium, which is radioactive—it emits so-called alpha particles in all directions. Geiger and Marsden let these alpha particles hit pieces of thin metal foil. To see where the particles went afterward, they made a special screen—a piece of glass coated with a chemical that would make a flash of light when it was hit by a particle.

By putting the screen in different places around the foil and counting the flashes, Geiger and Marsden could measure how many particles were coming from the foil in each of several directions. The results: most of the alpha particles passed right through the metal foil. But a small percentage bounced back.

Ernest Rutherford came up with an interpretation: the atoms in the foil were mostly empty space. But within each atom must be something hard and small and heavy—a nucleus. If a particle from the radium happened to hit a nucleus, it would bounce back.

So in 1911 Geiger, Marsden, and Rutherford shot tiny particles at some atoms, noticed that some of the particles bounced back, and from that concluded that an atom has a nucleus.

The following papers are reprinted in Henry A. Boorse and Lloyd Motz, eds., *The World of the Atom* (New York: Basic Books, 1966):

Hans Geiger and Ernest Marsden, "On a Diffuse Reflection of the Alpha Particles" (1909).

Ernest Rutherford, "The Scattering of Alpha and Beta Particles by Matter and the Structure of the Atom" (1911).

Hans Geiger and Ernest Marsden, "The Laws of Deflexion of Alpha Particles through Large Angles" (1913).

DEW-BOWS

You've probably looked for a rainbow in the sky after a thunderstorm on a warm afternoon. Try looking for a dew-bow in the grass on a cool, clear morning.

A dew-bow looks pretty much like a rainbow, but it's on the ground, and it's upside down compared to a rainbow. To see one you must have the sun at your back. Look toward the shadow of your head. The dew-bow—if it's there—forms part of a big circle around the shadow of your head.

The principle behind dew-bows and rainbows is the same. When sunlight hits a water droplet, it gets bent as it enters the water, then gets reflected from the back of the droplet, then gets bent again as it re-emerges into the air. The upshot is that most of the light comes back from a water droplet at an angle of about forty-two degrees from the direction it went in.

That forty-two-degree angle is characteristic of spherical water droplets. If rain and dew were made of some liquid other than water, the angle would be different.

The forty-two-degree figure is not exact. Different colors of light get bent at slightly different angles—that's why a rainbow or dew-bow separates white sunlight into colors.

On a bright, cool, dewy morning, the grass before you is covered with millions of water droplets, all illuminated by the sun. When you see a dew-bow—or a rainbow—you see light from only those droplets that happen to be in the right place to send sunlight back to your eye.

Marcel Minnaert, *The Nature of Light and Colour in the Open Air* (New York: Dover, 1954).

HALOGEN LAMPS

Ordinary light bulbs grow dimmer with age. As the light bulb burns, atoms evaporate from the hot tungsten filament and land on the inside of the glass. As more tungsten builds up on the inside of the bulb, less light gets out. Eventually the filament loses so much tungsten that it breaks—the bulb "burns out."

Halogen lamps are designed to fight this tungsten loss. The bulbs contain gas from the so-called halogen family of chemical elements. Iodine gas is frequently used. Inside a halogen lamp, iodine atoms combine with tungsten atoms evaporating from the filament to make a new molecule that won't stick to the inside of the bulb, as long as the inside of the bulb is very hot. Instead, this new iodine-tungsten molecule drifts to the filament and puts the tungsten atom back on it! The iodine atoms are then free to pick up more evaporated tungsten.

For this to work, the inside of the bulb has to be very hot—up to about

2,000 degrees Fahrenheit. The bulb of a halogen lamp is made small so it will be close to the filament and get hot. It's made of quartz instead of ordinary glass so it won't melt. And that tiny, delicate quartz bulb is often enclosed in a big glass bulb to protect it and to keep people's fingers off the hot quartz.

Because of their special construction, halogen lamps cost more but stay brighter and last longer than ordinary light bulbs.

H. R. Crane, "How Things Work: Halogen Lamps," *The Physics Teacher,* January 1985, pp. 41–42.

WHEN POP BOTTLES DON'T BLOW UP . . . AND WHEN THEY DO

Carbonated beverages in airtight bottles contain dissolved carbon dioxide gas ready to escape when the cap seal is broken. Let a factory-sealed bottle of carbonated beverage sit upright, quietly, all day, then carefully remove the cap. You hear a puff of escaping carbon dioxide, but the liquid stays put. On the other hand, everybody knows that if you shake the bottle for five seconds, then open it, you get a near-explosion with significant and lasting effects on nearby furniture and carpeting.

You might conclude that shaking the bottle increases the pressure inside. But it doesn't. Shaking does create bubbles. The beverage swirls and splashes and falls back on top of itself inside the shaking, sealed bottle. Here and there, carbon dioxide gas gets trapped below the liquid surface, making bubbles in the liquid that were not there before. The pressure is the same, but there are now more bubbles. When you open the bottle, those bubbles suddenly expand because the gas pressure inside them is higher than atmospheric pressure. The rapidly expanding bubbles push liquid out through the bottle neck.

If you don't shake the bottle, the liquid has few bubbles or none at all. Most of the carbon dioxide gas that's not dissolved invisibly in the liquid is above the liquid, and from there it can escape harmlessly when you loosen the cap.

However, a large and sudden temperature change can indeed alter the pressure inside a pop bottle and cause it to explode. Heating not only causes expansion of the carbon dioxide already present, but also makes more carbon dioxide come out of solution in the pop and rise to the surface as gas, raising the pressure even more.

So, as long as the temperature doesn't change, merely shaking a factory-sealed pop bottle will not increase the pressure inside—it just makes more bubbles. When you remove the cap, those expanding bubbles push liquid out of the bottle with a force whose ultimate application depends on your wisdom and judgment.

D. W. Deamer and B. K. Selinger, "Will That Pop Bottle Really Go Pop? An Equilibrium Question," *Journal of Chemical Education*, June 1988.

IT'S NOT WHAT YOU HEAR—IT'S WHEN YOU HEAR IT

You're sitting near the back of a big auditorium, and someone on the stage is talking. If you can hear the talker clearly, there's probably a sound-reinforcement system at work—a system involving microphones, amplifiers, and loudspeakers.

In some auditoriums, the talker's voice is fed not into one loudspeaker at the stage but into many loudspeakers mounted in the ceiling or along the side walls of the room. This approach puts a loudspeaker near everyone in the audience. But this approach creates a problem. Electronic signals can travel almost instantly to the back of a big room, while the live sound takes a noticeable fraction of a second to travel to the back through the air. So the best sound-reinforcement systems delay the electronic signal by a fraction of a second before sending it to loudspeakers at the back.

How long should that delay be? It might seem best for the amplified sound to reach you through the loudspeakers at exactly the moment live sound reaches you through the air. But audio engineers and psychologists have found that, for the most natural sound, the delay has to be just a little longer than that—maybe a fiftieth of a second longer.

This is because of a so-called precedence effect in human hearing. If the live sound reaches your ear about a fiftieth of a second before the amplified sound, then all the sound will appear to come from the stage, not from speakers in the ceiling. This is true even if the loudspeaker sound is somewhat louder than the live sound!

Sound appears to come from whichever source produces it first. If the delays in a sound-reinforcement system are set just right, you may never be aware that amplification is being used at all.

"Sound-Reinforcement System," in *McGraw-Hill Encyclopedia of Science and Technology* (New York: McGraw-Hill, 1987).

H. F. Olson, *Music, Physics and Engineering* (New York: Dover, 1967).

Weightless Water

Punch a hole somewhere near the bottom of an empty tin can. Fill the can with water. Of course, a stream of water squirts out of the hole. Now refill the container and drop it from a height of five or six feet. Notice that while the container is falling, water does not squirt out of the hole. The water is weightless with respect to the can as long as it is falling.

You can see how this happens by remembering that the force of gravity makes all falling objects—water, cans, and everything else—accelerate toward the ground at the same rate. (We're neglecting air resistance here, because it doesn't affect the main point.) When the punctured can sits on a table, it cannot accelerate toward the ground because it's not free to move. But the water presses against the sides of the container, escapes through the hole, and accelerates toward the ground.

When you drop the punctured can, gravity makes the can as well as the water accelerate toward the ground—at the same rate. Gravity pulls on the water, but it also pulls on the can. So the weight of the water does not press on the sides and bottom of the can. The water is weightless, and no longer squirts through the hole.

The water-filled can does not have to fall straight down for this demonstration to work. You might throw it sideways, for instance. However you throw it, once the can leaves your hand, no water squirts through the hole.

You might want to ponder this: if you could throw the can sideways at seventeen thousand miles an hour, it would be going as fast as an orbiting space shuttle. Again, the water inside would be weightless—and for the same reason that the shuttle astronauts are weightless in orbit.

D. R. Kutliroff, *101 Classroom Demonstrations and Experiments for Teaching Physics* (West Nyack, N.Y.: Parker, 1976).

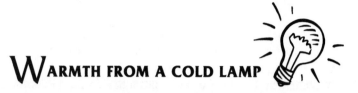

Warmth from a Cold Lamp

Find a lamp that has been turned off for several hours, so that the light bulb is stone cold. Touch the bulb with your fingers. It feels just about as warm or cool as any other glass object in the room.

Now, with your hand near the bulb but not touching it, switch the lamp on for five seconds, then switch it off. During the five seconds the lamp is on, you feel what seems to be heat coming from the bulb. Touch the bulb again immediately after you switch the lamp off; the bulb will feel almost as cool as it did before. It seems paradoxical—a light bulb that's not hot can make your hand feel warm.

What this experiment really demonstrates is that light and heat are not the same thing. When you turn on the bulb, what hits your hand is light—lots of visible light, and even more invisible infrared light. Light of any kind is really energy traveling though space. When light, visible or not, strikes matter—your hand, for instance—that energy can be converted to other forms. Here, the energy of light is converted to tiny random motions of the atoms in your skin: in other words, heat.

Heat can also be communicated directly by matter. If your hand is above the light bulb, you will soon receive additional heat from air, warmed by the bulb, rising to your hand. But you will feel warmth as a result of the light even if your hand is in the cool air below the bulb.

The sun warms the Earth by the same process: sunlight—visible and invisible—travels through 93 million miles of airless space, changing to heat only when it strikes the Earth or some other object, such as a spacecraft. It's true that the sun's surface is hot—like the filament inside the light bulb—but that heat is not directly communicated to faraway objects. The *light* from the sun is what warms the Earth.

Julius Sumner Miller, "Further Enchanting Things to Think About," *The Physics Teacher*, September 1979.

Mirror, Mirror on the Wall

Suppose you want a mirror to hang on the wall—a mirror big enough that you can see yourself full-length, from haircut to shoeshine, in one view. How big does that mirror have to be?

You might at first think that the mirror has to be as tall as you are. But that's not correct. If you hang it so the top edge is approximately even with the top of your head, the mirror has to be only about one-half your height to give you a full-length view of yourself, from head to foot.

To see why that's true, remember that looking into a mirror is something like looking through a hole in the wall. You can see big things through small openings. You can see a six-foot person through a keyhole if you get your eye close to the keyhole and if the six-foot person stands back from the other side.

Looking into a mirror is something like looking through a keyhole, with an important difference: the opening, represented by the mirror, is always exactly halfway between you and your reflection. If you stand three feet in front of a mirror, your reflected self appears three feet behind it. No matter where you stand, the mirror is always exactly halfway between you and your reflection. So looking at a mirror is like looking through a window located exactly halfway between you and your reflection. Because this window is halfway between you and your reflected self, it has to be only half as big as your reflected self to show all of your reflected self.

T. B. Greenslade, "Nineteenth Century Textbook Illustrations XXIV: The Half-Length Mirror," *The Physics Teacher*, September 1978.

Artificial Flavoring Means Fewer "Chemicals"

Chemists have found over one hundred different chemical compounds in ripe apples that contribute to their flavor. There are, of course, large amounts of sugars that give an apple its sweetness, and acids that add sourness. But the characteristic apple flavor comes from a balance of very small amounts of the rest of those hundred-plus different compounds. The flavors of other things we eat and drink arise from similar mixtures of many substances. In coffee, almost sixty compounds have been identified as contributing to the flavor; in Scotch whisky, over three hundred.

Just about all these compounds can be made in quantity in the laboratory, then mixed in the right proportions to reconstruct a particular flavor. The question facing a cost-conscious food processor is, how many of those dozens or hundreds of compounds are really necessary to make a convincing artificial flavoring? Just a few of the most important ones may be sufficient to simulate a natural flavor pretty well. Why make, say, a raspberry flavoring with one hundred compounds if a flavoring made with only ten comes close enough to the flavor of real raspberries?

Actually, one patented artificial raspberry flavor formula uses only seven different compounds. A patented strawberry formula calls for eight compounds; a patented chocolate formula uses just four. So, although we may tend to think of artificial flavoring as involving more "chemicals" than natural flavoring, the opposite is likely to be true. If an artificial flavoring tastes artificial, the reason may be that it contains not too many chemicals, but too few.

T. P. Coultate, *Food—The Chemistry of Its Components* (London: Royal Society of Chemistry, 1984). The raspberry formula (U.S. patent 3,886,289) is on p. 145.

CATCH A FALLING DOLLAR

I have here a crisp, straightened-out one-dollar bill. I hold it by a short edge between the thumb and forefinger of my left hand, allowing the bill to dangle flat and vertical. Now I place the thumb and forefinger of my other hand around the bottom edge of the hanging bill, not quite touching it.

I shall demonstrate my quick reaction time by releasing the bill with my left hand and catching it with my right hand before it has time to fall through my fingers. I've done some figuring, and I estimate that a dollar bill takes about a fifth of a second to fall a distance equal to its own length after I release it. I, however, am so quick that I can catch the bill before it has fallen even half its own length. There—I have caught the bill so that my thumb is over George Washington's portrait.

Now you try it. I'll dangle the bill so the bottom edge is between your fingers. Without warning, I drop the bill. You aren't quick enough to catch it. Again and again, you fail.

Actually I'm cheating. When I drop the bill and catch it myself, I do not demonstrate my own reaction time. My brain issues both the "drop" and "catch" instructions, so I am not really reacting to the motion of the bill. I can make the drop and the catch happen as close together in time as I want. I can even catch the bill before I drop it.

You, however, must see the bill begin to fall before doing anything. Messages must travel from your eyes to your brain to your hand—only then do your fingers close. All that takes time—more time than the dollar bill takes to fall a distance equal to its own length.

Do you think you'll react faster if I use a five-dollar bill instead of a one?

Martin Gardner, *Entertaining Science Experiments with Everyday Objects* (New York: Dover, 1981).

Cosmic rays

In 1912, the Austrian physicist Viktor Hess spent several nights hovering over Vienna in a balloon. During these flights, Hess measured how quickly a small piece of metal foil, charged with static electricity and sealed in an airtight can, would lose its electric charge at different altitudes.

Most of us have shuffled across a carpet on a cold, dry day to build up a personal static charge, and have felt that static electricity suddenly discharge in a miniature lightning bolt when we touch a faucet or something else connected to the ground.

Actually, any object charged with static electricity eventually loses its charge even if it's never grounded; the air always contains a few electrically charged atoms, or ions, that neutralize the static charge. Many of those ions are created by the Earth's natural mild radioactivity. Subatomic particles emitted by elements like uranium in the ground make ions by stripping electrons off atoms in the air. So measuring how quickly something loses static electricity, as Hess was doing, is really a way of measuring the general intensity of particles.

In 1912 it was generally thought that if you got higher above the ground, you'd detect fewer of these particles, because air between you and the radioactive ground would absorb some of the radiation. In his balloon flights, Hess found that radiation does taper off with altitude—but only up to a height of about one mile. Above one mile, Hess found increasing radiation—radiation coming from the sky.

Viktor Hess, in 1912, had discovered what are now called cosmic rays—fast subatomic particles from outer space. Physicists and astronomers now think that cosmic rays originate in exploding stars or bright centers of galaxies. But no one knows for sure.

V. F. Hess, "Penetrating Radiation in Seven Free Balloon Flights" (1912), trans. and reprinted in Henry A. Boorse and Lloyd Motz, eds., *The World of the Atom* (New York: Basic Books, 1966).

D. J. Helfand, "Fleet Messengers from the Cosmos," *Sky and Telescope*, March 1988.

CHOCOLATE BLOOM

Chocolate bloom is the gray film that sometimes appears on chocolate candy.

Most chocolate candy is about half cocoa butter, and there are six different forms of cocoa butter. They are all made of the same fat molecules, but their melting temperatures range from about 63 to about 97 degrees Fahrenheit. The difference is not in chemical composition, but in how the fat molecules are stacked up to make solid cocoa butter. Melting is basically the dismantling of the stacking pattern by heat. Some stacking patterns are more easily dismantled than others, so different forms of cocoa butter have different melting points.

Of these six different forms, only the fifth—the one that melts at about 93 degrees—has the nice glossy surface that everybody likes on chocolate candy. Much of the art of chocolate making involves getting the melted cocoa butter to solidify into that form and not into any of the other five forms.

If the chocolate isn't made correctly, or if it is subjected to a lot of temperature variations—making many trips into and out of the refrigerator—some of the fat molecules in the cocoa butter will be jostled out of their proper stacking pattern. Those molecules will emerge on the surface of the candy and settle into another stacking pattern—usually the one that melts at 97 degrees Fahrenheit rather than 93. That 97–degree form doesn't have a glossy surface.

So the gray film that sometimes develops on chocolate candy is made by fat molecules from cocoa butter separating from the chocolate and settling into a new stacking pattern with a dull surface.

Owen R. Fennema, ed., *Food Chemistry*, 2nd ed. (New York: M. Dekker, 1985).

T. P. Coultate, *Food—The Chemistry of Its Components* (London: Royal Society of Chemistry, 1984).

P. W. Atkins, *Molecules* (New York: Scientific American, 1987).

GETTING ROBBED IN THE CAMERA OBSCURA

"Camera obscura" literally means a dark room. For at least twenty-five hundred years, people have known that a single small hole in one wall of a closed, dark room will cause an image of the scene outside to be projected onto the opposite wall. A lens in the hole can brighten the image, but isn't absolutely necessary.

Since light travels in straight lines, each point in the projected image is formed by a ray of light coming through the hole from one point in the landscape outside.

The image projected by a hole in the wall is upside down. Light from the top of the landscape passes through the hole and ends up at the bottom of the projected image; light from the bottom of the landscape ends up at the top of the image. The ninth-century Chinese philosopher Shen Kua explained this with an elegant analogy: light rays going through a hole are constrained like an oar in a rowlock. When the handle of the oar is down, the blade is up, and vice versa. Some arrangement of lenses or mirrors can be added to make the image right side up.

A French writer of the seventeenth century tells about a camera obscura near a park reputed to be a hangout for young couples of low

136

morals. Fashionable patrons would pay to step into the camera obscura and watch projected images of various scandalous activities among the trees and bushes. The patrons were astonished not only by what they were seeing but by how they were seeing it—most of them didn't realize that because light travels in straight lines, a hole in a wall can project an image.

According to this seventeenth-century account, there were some other things these patrons didn't know: first, that most of the amorous action they were watching had been staged by the operators of the camera obscura, and second, that while they stood in the dark room, captivated by the projected image, their purses were being stolen!

J. H. Hammond, *The Camera Obscura: A Chronicle* (Bristol: Adam Hilger Ltd., 1981), p. 29.

CFCS AND CO$_2$

The news frequently mentions two gases in the atmosphere whose names sound similar but which are connected with different problems.

CFCs are chlorofluorocarbons, a family of man-made chemicals once used as propellants in spray cans and still used in air conditioners, refrigerators, and foam cushions. From the point of view of safety, CFCs at first glance seem ideal: they're non-toxic, non-flammable, and non-corrosive; they react chemically with very few substances.

But those desirable qualities are also the reason for the CFC problem. Chlorofluorocarbons can survive for decades in the lower atmosphere. That gives time for CFC molecules to rise to the upper atmosphere, where ultraviolet light from the sun strips chlorine atoms off the molecules. Those chlorine atoms break up molecules of the naturally occurring ozone in the upper atmosphere, in complicated reactions involving high-altitude clouds.

A molecule of ozone is made of three oxygen atoms bound together; a molecule of the oxygen we breathe has two oxygen atoms. The three-oxygen molecule blocks the sun's ultraviolet light, which would disrupt life on Earth if it reached the ground in large amounts.

In 1987 representatives of thirty-one countries, including the United

States, signed the so-called Montreal Protocol on Substances That Deplete the Ozone Layer, which includes guidelines for cutting back on CFCs during the 1990s. But some calculations indicate that even if chlorofluorocarbons are banned completely, the atmosphere may need a century to recover from damage already done.

CO_2, on the other hand, is carbon dioxide. CO_2 is exhaled by animals and people, and it's also generated by burning fossil fuels like petroleum, coal, and wood. Concern about carbon dioxide arises from observations that the percentage of carbon dioxide in the atmosphere has been increasing in the twentieth century, presumably because of human activity. Carbon dioxide traps infrared light generated by the Earth's warm surface—that's the so-called greenhouse effect.

Too much carbon dioxide in the atmosphere would trap too much infrared light and perhaps cause the world's climate to gradually warm up. Some studies indicate that the warming has already begun.

So CFCs and CO_2 have similar-sounding names, but are connected with two different atmospheric problems. CFCs, chlorofluorocarbons, are associated with depletion of the ozone layer; CO_2, carbon dioxide, is connected with the greenhouse effect.

R. Monastersky, many articles in *Science News*, including "The Decline of the CFC Empire," April 9, 1988; "Clouds without a Silver Lining," October 15, 1988; "New Chemical Model, New Ozone Fear," September 3, 1988; "Has the Greenhouse Taken Effect?" April 30, 1988.

R. A. Stolarski, "The Antarctic Ozone Hole," *Scientific American*, January 1988.

T. Beardsley, "Winds of Change: International Talks Address Human Effects on Climate," *Scientific American*, September 1988, pp. 18–19.

THE SHAPES OF SNOW

Snowflakes are ice crystals, and ice crystals can form as hexagonal plates, needles, or hexagonal columns, as well as the familiar starlike shapes.

All ice-crystal shapes are based in one way or another on the hexagon, the six-sided geometrical shape we also see in the cells of a honeycomb. The hexagon is basic to ice crystals because water molecules, when they link up to form ice, take positions corresponding to the corners of a hexagon.

Apparently, just about every snow crystal begins when a tiny amount of water freezes on a piece of dust high in the atmosphere. As this microscopic particle drifts through a cloud, it picks up more and more water molecules from the surrounding humid air. Each new molecule hooks into the existing hexagonal pattern of the crystal. Eventually this process makes an ice crystal, big enough to see, with some kind of six-sided symmetry.

Meteorologists have found that the final shape of an ice crystal is very sensitive to the temperature at which it forms. Below about minus 10 degrees Fahrenheit, ice crystals form as hollow hexagonal columns—something like the shape of a pencil. Up to about 3 degrees Fahrenheit above zero, ice crystals form hexagonal plates; between 3 and 10 above, the branching star-shaped types, the so-called dendrites, appear. In warmer air, ice crystals come out as plates, needles, or solid hexagonal columns, depending on the exact temperature.

If an ice crystal drifts through several different temperature regions within a cloud as it forms, it may come out as a hybrid. One of the most beautiful hybrid types is a hexagonal column with a flat plate at each end. These have been named tzuzumi crystals, after the Japanese drums they resemble.

Charles and Nancy Knight, "Snow Crystals," *Scientific American*, January 1973.
B. J. Mason, "The Growth of Snow Crystals," *Scientific American*, January 1961.
(The previous two articles are reprinted in the *Scientific American* anthology *Atmospheric Phenomena*, intro. by David K. Lynch [New York: W. H. Freeman and Co., 1980].)
Donald Ahrens, *Meteorology Today*, 3rd ed. (St. Paul: West, 1987).

SKIDDING DOWNHILL

Let a toy car roll down a long, steep wooden ramp. If all four wheels turn freely, the car will be going pretty fast when it gets to the bottom.

Now lock all four wheels, maybe with tape stuck to the wheels across the fenders. The car now skids, out of control, down the ramp.

What happens if you lock only the rear wheels? You might guess that the car will now go down the ramp front end first, since the rear wheels skid while the front wheels roll, and friction from the skidding wheels pulls the rear end back—right?

Well, try it—lock only the rear wheels, leave the front wheels free to turn, point the toy car down the ramp, and let it go. You'll probably see an amazing result: the car spins around and goes down the ramp backwards—that is, with the locked wheels ahead and the freely rolling wheels trailing behind!

Rolling wheels have better contact with the road than skidding wheels because the bottom point of a rolling wheel is always at rest on the road. Also, it's always harder to start something sliding, or skidding, over a surface than it is to keep something sliding once it's started: static friction is greater than sliding friction.

So, as the car goes down the ramp, static friction under the front wheels maintains good contact with the ramp. The front end becomes a pivot. Meanwhile, the much smaller sliding friction under the locked rear wheels allows the rear end to come around to the front.

This demonstrates one safety advantage of anti-lock brakes on a real car. As long as the wheels roll rather than skid, you're using static friction rather than sliding friction to bring your car to a stop. Also, you can steer better when you're rolling than when you're skidding.

INFRARED

Around 1800 the German-born English astronomer William Herschel decided to study the sun. He figured that since the sun is the star nearest to the Earth, if he could learn something about the sun he might understand something about the other stars.

Some of Herschel's conclusions turned out to be wrong—such as his idea that sunspots are mountains. But Herschel was right when he concluded from his experiments that there is more to sunlight than meets the eye. He allowed sunlight to pass through "various combinations of differently-coloured darkening glasses." Herschel wrote: "What appeared remarkable was, that when I used some of [these darkening glasses], I felt a sensation of heat, though I had but little light; while others gave me much light, with scarce any sensation of heat." Sunlight apparently contained invisible rays of heat as well as visible rays of light.

Herschel let sunlight pass through a prism to separate it into a spectrum of rainbow colors and placed a thermometer in different parts of the spectrum to see how much heat it picked up. Rays carrying the most heat were beyond the red end of the spectrum, where almost no light could be seen. Soon Herschel demonstrated that these rays beyond the red end of the spectrum could be bent and focused by prisms, mirrors, and lenses, just like ordinary visible light.

William Herschel, in the year 1800, had discovered a form of light not visible to the eye—what is now called infrared light. Hot objects like the surface of the sun and incandescent light bulbs give off a lot of infrared light.

"William Herschel," in *Dictionary of Scientific Biography* (New York: Charles Scribner's Sons, 1972).

Dark Meat and Light Meat

Chickens and turkeys have dark, and sometimes greasy, leg muscles and light, dry breast muscles. On the other hand, flying game birds like ducks and geese have dark breast muscles. These differences are connected with how different muscles get energy.

The dark color of dark meat comes not so much from blood as from a special protein that stores oxygen in muscles that move the skeleton. This protein goes by the name of myoglobin. When a myoglobin molecule is loaded with oxygen, it's deep red in color. When it loses oxygen, it becomes pale purple. When it's cooked, myoglobin turns brown.

Muscle fibers rich in myoglobin fall into the broad category of slow-twitch muscle fibers. They tend to be especially well suited to slow or sustained activity, like walking around a barnyard. Muscles that store oxygen in myoglobin and fuel in the form of fat can operate steadily even when the oxygen supply from the blood is low. So dark meat tends to come from muscles that are used constantly—for instance, from the leg muscles of chickens and turkeys and the breast muscles that operate the wings of ducks and geese.

The other broad category of muscle fibers is the fast-twitch category, suited for brief spurts of high power. Fast-twitch muscle fibers store less oxygen than slow-twitch fibers because they contain less myoglobin. So fast-twitch fibers tend to be light-colored. They burn not fat but carbohydrates—mainly glucose brought in by the bloodstream. Oxygen to burn the glucose must also come from the bloodstream. High-powered spurts of activity have to be brief because fast-twitch muscle fibers can operate for only a short time after they use up the available oxygen supply.

P. W. Atkins, *Molecules* (New York: Scientific American, 1987).
Helena Curtis, *Biology*, 4th ed. (New York: Worth, 1983).
Harold McGee, *On Food and Cooking: The Science and Lore of the Kitchen* (New York: Macmillan, 1987).

SPERM MEETS EGG: WHAT REALLY HAPPENS

In the 1870s biologists knew that, as a rule, animal egg cells would start to develop only when they had been "fertilized"—whatever that meant. But good microscopes and new techniques of that era made it possible to see what really happens.

In 1874 the German zoologist Oscar Hertwig studied fertilization of sea urchin eggs. Sea urchins are relatives of starfish and sand dollars, abundant near seashores, and still favorites among biologists. Fertilization of sea urchin eggs is easy to observe because it happens in open water.

Hertwig saw that before fertilization the sea urchin egg had a single nucleus, a distinct body inside the cell that divides when the cell divides; but immediately after fertilization there were not one but two nuclei, which soon fused into one. Hertwig concluded in 1875 that the extra nucleus had been added to the egg by a sperm cell. At about the same time a French biologist, Hermann Fol, actually saw a sea urchin sperm cell penetrate an egg cell.

By about 1900 the observations of Hertwig, Fol, and others were brought together to make the now-familiar picture: a single sperm cell penetrates the egg cell; the nucleus of the sperm merges with the nucleus of the egg; then development begins.

These observations suggested but did not yet prove that the sperm contributes something essential to the new individual. Not till the twentieth century was it demonstrated that one function of a sperm cell is to add genetic information to the nucleus of the egg. The other function of the sperm is to trigger development of the egg into an embryo.

Garland E. Allen, "Hermann Fol," and Robert Olby, "Wilhelm August Oscar Hertwig," in *Dictionary of Scientific Biography* (New York: Charles Scribner's Sons, 1971, 1972).
Oscar Hertwig, *Allgemeine Biologie*, 3rd ed. (Jena: G. Fischer, 1909), pp. 292ff.

Do you *really* have a one-track mind?

This is a psychology demonstration in which I ask you, my volunteer subject, to perform three tasks.

First, I show you a list of names of colors—"red," "blue," "green," "orange," and so on—written in clear black letters on white paper. I ask you to read the list aloud as fast as possible. You have no trouble with this.

For your second task, I show you another list of color names, which I have written not in black ink but in ink of various colors, using a different felt-tip marker for each color name. Again I ask you to read the list aloud as fast as possible, and again you have no trouble.

For your third and final task, we use the colored list again. I ask you not to read the color names on the list, but to tell me as fast as you can what color of ink I used to write each color name. Now you have trouble! You see that I've used each marker to write the name of a color different from the color of ink in the marker. I've written the word "red" in blue ink, the word "yellow" in green ink, and so on.

When you look at the word "yellow" written in green ink, it's hard not to blurt out "yellow" instead of naming the color of the ink as I asked you to do. You have to resist the tendency to read.

What this shows is that our perceptual system has one part that reads and another part that judges color. This test, in which the name of one color is written in ink of another color, forces both those parts of the perceptual system to generate color names at the same time. That rarely happens in everyday life.

J. R. Stroop, "Studies of Interference in Serial Verbal Reactions," *Journal of Experimental Psychology* 18:643–662 (1935).

H. R. Schiffman, *Sensation and Perception: An Integrated Approach* (New York: Wiley, 1982).

The season of static electricity

As everybody knows, you can acquire a personal charge of static electricity by shuffling across a carpet on a dry winter day. When you do this, atoms in your shoes give up a few electrons to atoms in the carpet. You may be surprised to hear that the exact details of how that happens are not completely known. But the result is that you are left with a slight electron deficit relative to other objects, while the carpet ends up with a slight electron surplus. In other words, you and the carpet acquire equal and opposite static charges.

Now bring your finger near a faucet or another person, and electrons will jump across the gap to your finger, wiping out your electron deficit in a miniature lightning bolt.

This kind of thing happens most often on cold, dry winter days, because on those days indoor relative humidity is likely to be low. In dry air, objects with surplus electrons keep them more easily, and objects short on electrons can't pick up electrons easily through the air. In other words, dry air is a good insulator. Humid air, on the other hand, has a slight ability to conduct electricity; that allows static charges to drain away.

Next question: why is indoor air in winter so often dry? Remember that the amount of water vapor cold outdoor air can hold is smaller than the amount that same air can hold when it's warmer. Bring cold air in from outside, heat it up, and you lower its relative humidity—not by removing water vapor, but by increasing the air's capacity for water vapor.

So you're most likely to see effects of static electricity indoors on cold days, because indoor air on those days is likely to have low relative humidity, and because dry air is a good insulator.

A. D. Moore, "Electrostatics," *Scientific American*, March 1972.

A. D. Moore, ed., *Electrostatics and Its Applications* (New York: Wiley, 1973).

D. S. Ainslie, "What Are the Essential Conditions for Electrification by Rubbing?" *American Journal of Physics* 35:535–537 (1967).

Richard P. Feynman, *The Feynman Lectures on Physics* (Reading, Mass.: Addison-Wesley, 1964), vol. II, chap. 1.

WHAT SOAP DOES TO WATER

Water molecules attract each other. The molecules at the surface of a body of water make a film under tension. That film is strong enough to support a needle or a small insect like a water strider. Surface tension also pulls water into round droplets. Soap breaks surface tension, and here's how to see it happen.

144

You need a clean dinner plate, some talcum powder, and a bar of soap. Rinse the plate thoroughly to get rid of any grease or soap. Fill the plate with water and sprinkle a trace of talcum powder on the surface.

Now take the bar of soap and touch one corner of it to the water surface near the edge of the plate. The talcum powder will be pulled suddenly to the opposite side of the plate.

The talcum powder doesn't dissolve in the water and it doesn't sink; it lies on the water surface, supported by surface tension. The effect of the soap is to break that film of surface tension. At the place where soap touches water, molecules of water attract molecules of soap rather than each other. Meanwhile, on the other side of the plate, water molecules still attract each other. Touching the soap to the water is like cutting a stretched rubber band. The surface tension on the far edge of the plate yanks the talcum powder away from the soap.

C. J. Lynde, *Science Experiences with Home Equipment*, 2nd ed. (Princeton, N.J.: Van Nostrand, 1949).

ROLL OVER, GEORGE WASHINGTON

Put a quarter on the table, with George Washington's head right side up. Hold the quarter down with your finger. Now put another quarter flat on the table, again with Washington right side up, so the edge of the second coin touches the twelve o'clock position on the edge of the first coin. If you roll the second coin halfway around the edge of the first coin, from the twelve o'clock position to the six o'clock position, will George Washington come out right side up or upside down?

As you roll from twelve o'clock to six o'clock, exactly half the circumference of the rolling coin touches the edge of the coin you're holding down. So it seems that the rolling coin goes through half a turn, and Washington should come out upside down. But when you try it, George Washington comes out right side up!

How can that be? If you roll a twenty-five-cent piece along a straight edge, like a ruler, for a distance equal to half its circumference, Washington comes out upside down, because the coin rolls half a turn in that situation. But when you roll one twenty-five-cent piece halfway around another, the curvature of the edge of the coin you're holding down with your finger adds another half-turn. The two half-turns add up. The rolling coin in our demonstration actually goes through one full turn.

And that's how George Washington comes out right side up.

R. J. Brown, *333 Science Tricks and Experiments* (Blue Ridge Summit, Pa.: Tab, 1981).

Martin Gardner, *Entertaining Science Experiments with Everyday Objects* (New York: Dover, 1981).

Don't believe your fingers

Most of the illusions we hear about and enjoy playing with are optical illusions. This is a tactile illusion.

To experience this illusion, all you need is a marble or a pea. Cross your fingers, extending the middle finger over the index finger so the two fingertips are next to each other, but reversed from their normal arrangement. Now roll the marble around on a table with your crossed fingertips. Almost immediately you'll probably get the distinct impression that there are two marbles, not just one.

Crossing your fingers makes information travel to your brain through unusual channels. Normally the outside of your middle finger and the inside of your index finger face away from each other. Crossing your fingers brings those two sides together. When the marble gets between your crossed fingertips, it touches areas that normally would be touched only if there were two marbles. Cross your fingers, and one marble feels like two.

This effect has been known long enough to be called Aristotle's illusion. Here are some modern variations.

Try touching your nose instead of a marble. Your fingertips may give you the impression that you have two noses.

Try different fingers: cross your middle finger over your ring finger. Then lay a pencil over the two crossed fingertips so the barrel of the pencil touches one fingertip and the point touches the other. See if you can tell, just by touch, which way the pencil is pointing. Rocking the pencil gently back and forth may enhance the strangeness and vividness of the sensation.

W. James, *Principles of Psychology*, vol. 2 (New York: H. Holt, 1890), p. 86, "Illusions of the First Type."

THE CURVE OF A MEANDERING RIVER

It's a peculiar kind of curve. You can spot it easily on a map anywhere it appears—for example, along the Mississippi River between Arkansas and Tennessee. The curve is not a series of pieces of a circle connected together. And it's not one of the familiar curves used in science and engineering to describe waves. It's something different.

It comes about because a river, in a manner of speaking, doesn't like to make sharp turns. A sharp turn offers resistance to the flow of the water. A river seeks the path that distributes that resistance to flow as evenly as possible along the length of the river. So the path is not jagged like the letter W, but curved in a particular way something like the letter S.

You can see that curve elsewhere—for instance, in the hem of a pleated curtain. The pleats at the top of the curtain try to bend the hem at the bottom into some kind of zigzag, but the fabric resists bending and makes a smooth, even curve like the curve of a meandering river. The ruffles on a ruffled blouse and the loops of a ribbon bow make versions of that meandering-river curve.

What all these examples have in common is that whatever it is that's being bent offers a uniform resistance to bending—and that determines the final shape.

Luna Leopold and W. B. Langbein, "River Meanders," *Scientific American*, June 1966 (reprinted in the *Scientific American* book *The Physics of Everyday Phenomena*, ed. Jearl Walker [San Francisco: W. H. Freeman, 1979]).
John S. Shelton, *Geology Illustrated* (San Francisco: W. H. Freeman, 1966).

CURVED SPACE IN A CHRISTMAS ORNAMENT

The Christmas ornaments we're talking about here are those glass spheres with a mirror-like reflecting surface. Actually, for this observation you don't need a Christmas ornament; any ball-shaped object with a surface that reflects like a mirror will do.

As you look at your reflecting ball, you see the reflection of your own face at the center. Around your head you see the whole room—walls, floor, and ceiling. The farther you look from the center of the ball, the more the image is distorted. That's why your nose looks extra large: the rest of your face is smaller because of the distortion.

As you look more closely, you see that the ball reflects almost the whole world as seen from where the ball is. The only part left out is the tiny piece of the scene right behind the ball, as you view it.

Now suppose someone behind you spreads a square red tablecloth on the floor and measures the edge of it with a ruler. The reflected tablecloth certainly isn't square; the reflection is distorted. But the reflected ruler still shows all four sides to have the same length, because the ruler is distorted too.

The distorted reflection in a shiny ball is a model of curved space, one of the more esoteric concepts of mathematics. People living in a curved space may not know that they live in a curved space, because their rulers and measuring instruments are curved just like everything else.

Marcel Minnaert, *The Nature of Light and Colour in the Open Air* (New York: Dover, 1954), p. 13.

Guests in every cell

Mitochondria are the indispensable power plants of living cells, including the cells we're made of. They convert the stored energy of carbohydrates from food into other forms the cell can use. Mitochondria use oxygen to do this. We need to inhale oxygen because our mitochondria need it.

In photographs taken with electron microscopes, mitochondria often appear as potato-shaped things inside a cell. A cell may contain anywhere from a few to a few thousand mitochondria.

In some ways mitochondria resemble bacteria. Inside their host cell, mitochondria reproduce by dividing, just as bacteria do. Each mitochondrion has its own DNA molecule to store genetic information, and that molecule is in the form of a loop, just like the DNA in bacteria. Other striking chemical similarities between mitochondria and bacteria have been found. Those resemblances point to the conclusion that today's mitochondria are really descendants of free-living bacteria of long ago.

The presumed ancestors of mitochondria might have distinguished themselves a billion or so years ago by (so to speak) mastering the trick of using oxygen to manage energy. Back then, oxygen in the air was a new thing; other types of bacteria were starting to make oxygen as a byproduct of photosynthesis. Somewhere, a bacterium of the oxygen-using type took up residence inside a larger cell and, in so doing, conferred upon its host the ability to thrive in an oxygen-rich atmosphere. That host cell then became the ancestor of today's plants and animals, including us.

So today we have guests in every cell: mitochondria, distant descendants of free-living bacteria of a billion years ago.

L. Margulis, "Symbiosis and Evolution," *Scientific American*, August 1971 (reprinted in the *Scientific American* anthology *Life: Origin and Evolution* [San Francisco: W. H. Freeman, 1979]).

The Glory

Next time you fly, try to get a window seat on the shady side of the plane, not above the wing. You want a clear view of the airplane's shadow on clouds below. Look for a bull's-eye pattern of bright rings, tinged with rainbow color, surrounding the shadow on the clouds: the so-called glory.

The glory is not the same thing as a rainbow. A rainbow is made of sunlight coming back from water droplets at an angle of about forty-two degrees from the direction it went in. The glory, on the other hand, is light coming out of the droplet in a direction almost exactly opposite to the incoming sunlight. Physicists have found the explanation extremely complicated, but part of the story goes like this:

Light encountering a spherical water droplet in a cloud has, so to speak, several options. One is to enter one side of the water droplet at a glancing angle, bounce once off the inner surface of the back of the droplet, then exit the droplet on the other side—basically making a U-turn. Another option for a light ray is to bounce not once but fourteen times off the inside of the droplet, making three and a half trips around the droplet in the process (like a car searching a lot for a parking space), then to exit the droplet.

It turns out that light that has bounced once and light that has bounced fourteen times will emerge from the droplet going in almost exactly the same direction. A doubly strong light ray goes almost straight back toward the sun, and hence, toward your airplane. But this complicated process also breaks white light into its colors, and different colors emerge in slightly different directions. That's why the glory is tinged with color.

So, if you have a window seat on the shady side of a plane flying not too high over clouds, look for the glory: a mysterious bull's-eye pattern of colored rings surrounding the shadow of the airplane.

H. C. Bryant and N. Jarmie, "The Glory," *Scientific American*, July 1974 (reprinted in the *Scientific American* book *Atmospheric Phenomena*, intro. by D. K. Lynch [New York: W. H. Freeman and Co., 1980]).

THE WATERFALL EFFECT

Sit next to a stream and stare at a waterfall. After a few minutes of watching the falling water, look away. For a few seconds you get the impression that everything else is rising!

Psychologists who have studied this so-called waterfall effect have found evidence of separate detectors in our visual system for upward motion and downward motion. Even when we look at something that's not moving, the "up" and "down" detectors are both active. But they are equally active, so their signals balance out to give us an impression of no motion.

Staring at a waterfall apparently tires the downward-motion detectors, so they become less active. Meanwhile, our upward-motion detectors continue to send the same signal as for a stationary object. When we look away from the waterfall, we get the "up" signal, but without much of a "down" signal to balance it. So the "up" signal dominates and gives us the impression that a stationary landscape is rising.

Watch for a similar effect when you're riding as a passenger in a car or a bus. Look at the landscape going by the window. Stare at some object on the ground in the middle distance. Your motion makes the landscape appear to rotate around the point you're staring at. Things closer to you appear to move backward; things farther away appear to move forward. As you shift your gaze from point to point, the landscape appears to rotate in the same direction around each new point. All this stimulates, and soon tires, your detectors for that direction of rotation.

Now, what happens when your vehicle stops? For a few seconds those fatigued rotation detectors can't balance the signal coming from detectors sensitive to rotation in the other direction. When you stop, the landscape appears for a moment to rotate the wrong way!

O. E. Favreau and M. C. Corballis, "Negative Aftereffects in Visual Perception," *Scientific American*, December 1976.

Marcel Minnaert, *The Nature of Light and Colour in the Open Air* (New York: Dover, 1954).

Philip G. Zimbardo, *Psychology and Life*, 11th ed. (Glenview, Ill.: Scott, Foresman, 1985).

White sky, gray snow

If you go outside on a completely overcast day when the ground is covered with snow, you get the impression that the sky is gray and the snowy ground is white—brighter than the sky. But that impression is wrong. If you don't believe it, prop a hand mirror in the snow so it reflects the cloudy sky, then step back and look at the mirror in comparison to the snow. The overcast sky, relected in the mirror, will look brighter than the snow.

The snow cannot possibly reflect more light than it receives from the sky. The snow may be almost as bright as the sky, but it cannot be brighter. The apparent whiteness of the snow is a psychological effect, perhaps generated by contrast between the snow and dark objects like trees. (We're assuming here that the sky is uniformly overcast, absolutely featureless, with no difference in sky brightness from one place to another.)

You may have noticed the effect we're talking about if you've ever taken a picture of a white house on an overcast day. You know the house is white, but it looks disappointingly gray in the photograph. The camera records the sky as bright and the house as less bright, the way they really are.

A photograph taken on a cloudy day does not match the impression given by the eye because a camera does not have the complex system for interpreting images that our visual system has.

Marcel Minnaert, *The Nature of Light and Colour in the Open Air* (New York: Dover, 1954).

WHY CHICKENS DON'T HAVE WEBBED FEET

In a certain embryonic stage, chickens have at least the beginnings of webbed feet. Embryologists have observed that as toes develop in a chick embryo's foot, cells grow between the toes. The same thing happens in the embryonic foot of a duck.

But one of the differences between a chicken and a duck is that in a duck most of those cells between the toes survive to become webbing. In the chicken embryo those cells die in a later stage of development, leaving the chicken with separated toes. In other words, a chicken's foot is sculpted not only by genetically programmed formation of new cells, but also by genetically programmed death of existing cells.

This doesn't seem to be the most efficient way to make a chicken's foot. How did it come about that a chick embryo makes cells between its toes, only to kill them off a short time later? This extra step seems to indicate that chickens and ducks have both inherited a plan of development, a schedule for getting the right cells into the right places at the right times, from some common ancestor. The chicken's foot represents a modification of that original plan. The result of all this is that ducks have feet suited for swimming, and chickens have feet suited for walking.

Genetically programmed death of certain cells has been seen in other situations too. In newborn mammals, many muscle fibers have more than one connection to the nervous system. But in the first few weeks of life, most of the original nerve-to-muscle connections die off, leaving only one nerve fiber connected to each muscle fiber.

Death of specific cells may also be the process that separates the two

152

parallel bones in our forearms, the ulna and the radius, when our skeletons are developing.

Scott F. Gilbert, *Developmental Biology*, 2nd ed. (Sunderland, Mass.: Sinauer Associates, 1988).

Can you draw a penny from memory?

There are eight important features on a U.S. penny: a head; "In God We Trust"; "Liberty"; a date; since 1959, a building; "United States of America"; "E Pluribus Unum"; and "One Cent." Now, can you draw both sides of a penny from memory, putting each feature in the right place, with the right orientation?

Can you remember whose head is on the penny? Which way is the head facing? What is the building? Does the building have columns, or a dome, or chimneys?

Two psychologists, Raymond Nickerson and Marilyn Adams, asked a group of adult U.S. citizens to draw a penny from memory. Even when given that list of eight features to include in their drawings, no one in a group of twenty people located all eight features correctly.

Of course, in real life we need to recognize pennies, not draw them from memory. So Nickerson and Adams did another experiment. They showed people fifteen drawings of a penny, all but one of which had some feature missing or misplaced. Only fifteen out of thirty-six people could pick out the correct drawing, and those people didn't indicate a very high degree of confidence.

Why do we have such a poor memory for a common object? Psychologists are working on this question, partly because it bears on the important practical issue of the reliability of eyewitness testimony. Maybe we remember just enough to distinguish a penny from other coins.

Another factor impairing our memory of a penny may be interference from memories of other coins. Have you ever noticed that the head on a penny faces the opposite direction from heads on other U.S. coins? If your drawing of a penny had the head facing the wrong way, maybe you were in some way remembering a nickel or a dime.

R. S. Nickerson and M. J. Adams, "Long-Term Memory for a Common Object," *Cognitive Psychology* 11:287–307 (1979).

E. Loftus, *Memory: Surprising New Insights into How We Remember and Why We Forget* (Reading, Mass.: Addison-Wesley, 1980).

Warren E. Leary, "Novel Methods Unlock Witnesses' Memories," *New York Times*, Tuesday, November 15, 1988.

Clouds in a Jar

In 1894 the English physicist Charles Thomson Rees Wilson stood on a hilltop in Scotland, watching the play of sunlight on the clouds below. Wilson saw and remembered the glory, a bull's-eye pattern of colored rings of light that surrounds shadows on clouds.

Back at his lab, Wilson tried to make artificial clouds in a jar in order to study the glory more closely. The basic idea was to seal humid air in a jar, then cool the air suddenly by making it expand, using a hydraulic gadget. Wilson's cloud chamber worked: when the air was expanded suddenly by the right amount, dense fog appeared in the jar.

Wilson was imitating what happens in the atmosphere: humid air rises, expands, and cools, just like air escaping from a high-pressure nozzle. Cool air cannot hold as much water vapor as warm air, so conditions are right for water vapor to condense into cloud droplets.

Wilson knew that a cloud droplet is most likely to form if there's something for the water vapor to condense on—a tiny piece of dust, or even an electrically charged air molecule. This gave him an idea.

In 1911 Wilson put radioactive material near his cloud chamber and saw not a fog but "little wisps and threads of cloud" in the jar. He realized that those wisps and threads of cloud were tracks left by subatomic particles from the radioactive stuff, flying through the jar, knocking electrons off air molecules in their path, leaving those molecules electrically charged. Cloud droplets formed first around the charged air molecules.

Wilson's cloud chamber, invented for meteorology, turned out to reveal tracks of particles smaller than the atom, and soon became one of the most important tools in nuclear physics.

"Charles Thomson Rees Wilson," in *Dictionary of Scientific Biography* (New York: Charles Scribner's Sons, 1976).

C. T. R. Wilson, "On an Expansion Apparatus for Making Visible the Tracks of Ionising Particles in Gases and Some Results Obtained by Its Use" (1912), reprinted in Henry A. Boorse and Lloyd Motz, eds., *The World of the Atom* (New York: Basic Books, 1966).

Ultraviolet

Let sunlight pass through a prism, let the emerging rays strike a screen, and you get a spectrum of rainbow colors. The order of colors is always the same: red, orange, yellow, green, blue, violet. At each end of the spectrum, light fades into invisibility.

In 1800 the astronomer William Herschel discovered that sunlight contains invisible rays beyond the red end of the spectrum—rays now called infrared light. News of the discovery quickly reached a German philosopher, Johann Wilhelm Ritter, who believed in the Romantic idea that the world was built on a principle of polarity: positive and negative electricity, north and south magnetic poles, and so on. Ritter guessed that if there were invisible rays beyond the red end of the solar spectrum, there might also be invisible rays beyond the other end—the violet end.

Ritter looked for these rays by using paper soaked in a chemical that would turn black when exposed to light—a primitive photographic emulsion. He found the greatest blackening just beyond the violet end of the spectrum, where no light was visible to the eye. Ritter, in 1801, had discovered what is now called ultraviolet light.

Johann Wilhelm Ritter seems to have been inspired by an individual philosophical belief. But his discovery was scientific in that anyone, of any philosophical bent, could repeat his experiment and see that there are invisible rays, ultraviolet rays, beyond the violet end of the spectrum.

R. J. McRae, "Johann Wilhelm Ritter," in *Dictionary of Scientific Biography* (New York: Charles Scribner's Sons, 1975).

Reproduction and Sex: What's the Difference?

Some living things reproduce without sex. An amoeba, for instance, is a single-celled organism that simply divides to make two individuals out of one. As the biologists Rollin Hotchkiss and Esther Weiss once wrote, if humans reproduced that way, then each of us, at about age twenty-five, would abruptly divide into two identical people; twenty-five years later, each of those people would divide again, and so on.

Sex, in the technical sense, is the process in which genes from two individuals are combined in new arrangements—like dealing a new hand of cards. In some living things there is sex without reproduction. The single-celled paramecium, for example, joins with another of its kind, exchanges genes with its partner, then separates with a new genetic identity. The process makes no new individuals, but it does create variations among existing paramecia that help the species to adapt and survive. That seems to be the real advantage of sex.

If people were like paramecia, sex would not result in pregnancy, but would alter the eye color and the height of both partners! Of course, this is absurd and impossible because people are made of so many cells. There's no way to make identical genetic alterations in all our cells at once.

In humans and other organisms made of many cells, the combining of genes—the dealing of the new hand of cards—happens only when sperm and egg meet. Development of the egg into an embryo then begins immediately. For us, sex and reproduction are inextricably linked.

R. D. Hotchkiss and E. Weiss, "Transformed Bacteria," *Scientific American,* November 1956 (reprinted in the *Scientific American* anthology *Genetics,* intro. by C. I. Davern [San Francisco: W. H. Freeman, 1981]).

Scott F. Gilbert, *Developmental Biology* (Sunderland, Mass.: Sinauer Associates, 1985).

A COMB, A SOCK, AND A FAUCET

A comb and a wool sock can demonstrate the structure of a water molecule.

On a day of cold, dry weather, turn on the bathroom faucet just enough to get a stream of water the diameter of a pencil lead. Now rub a comb on a wool sock. Slowly bring the teeth of the comb near the water stream. If conditions are right, the comb attracts the water. You may be able to bend the water stream through an angle of ninety degrees or even more.

This happens because you've given the comb a charge of static electricity by rubbing it on the sock. But why should a charged comb attract water? It happens because water molecules are asymmetrical; one side of each water molecule has a positive electrical charge, the other side a negative charge.

Remember that in the world of electricity, like charges repel and opposites attract. If the comb has a negative charge, it'll attract the positive side of each water molecule and repel the negative side. So water molecules in the stream will turn around so their positive sides are nearest the comb.

But there's more. The electric force field gets stronger as you get closer to the tip of each comb tooth. Each water molecule's positive side is a little closer to the comb than its negative side. So the attraction between the comb and the positive side of the water molecule is a little stronger than the repulsion between the comb and the negative side of the same molecule. The attractive force therefore wins out, and the water molecule is drawn toward the comb.

So a comb, charged with static electricity by being rubbed on a wool sock, will attract a stream of water. Notice that the sock will attract the stream, too, because the rubbing gives the comb and the sock charges that are opposite but equal.

Many similar experiments are described in C. J. Lynde, *Science Experiences with Ten-Cent Store Equipment* (Scranton, Pa.: International Textbook Co., 1951).

A. V. Baez, "Some Observations on the Electrostatic Attraction of a Stream of Water" (letter to the editor), *American Journal of Physics* 20:520 (1952).

R. Gardner, *Ideas for Science Projects* (New York: F. Watts, 1986).

Our MOST DISTANT RELATIVES

Are we humans more closely related to earthworms or to jellyfish? The answer is not obvious—we don't look like either one.

Biologists have developed a molecular method to find out about distant relationships among living things. The idea is to look not at body structure but at the molecular structure of ribosomal ribonucleic acid—"ribosomal RNA" for short. Every living cell has ribosomal RNA; it's indispensable for executing the cell's genetic instructions.

A ribosomal RNA molecule is a long chain of thousands of smaller molecules, hooked up like cars in a train. Those smaller molecules come in only four types and are connected in a very specific order in each species. In other words, each ribosomal RNA molecule is like a long coded message spelled in a four-letter alphabet.

Now there are laboratory techniques to determine the spelling of the message, the order of smaller units in the ribosomal RNA molecule. It turns out that the spelling of this message is slightly different from one species to another. Species that spell the message almost exactly the same way are almost certainly more closely related than species whose spellings differ more.

This molecular method indicates, among other things, that we humans are in fact more closely related to earthworms than to jellyfish. The sequence of smaller units in our ribosomal RNA is closer to the sequence in earthworms than to the sequence in jellyfish.

This implies, in turn, that humans and earthworms had a common ancestor, long ago—and that humans, earthworms, and jellyfish had a common ancestor even farther back. Today's molecular similarities record yesterday's family trees.

K. G. Field et al., "Molecular Phylogeny of the Animal Kingdom," *Science* 239:748–753 (February 12, 1988).

CHEAP IMITATION IVORY

In 1863 the American firm of Phelan and Collander, manufacturer of traditional ivory billiard balls, offered ten thousand dollars to anyone who could develop a substitute for natural ivory.

The American printer John Wesley Hyatt responded in 1868 with a material he called celluloid. It was one of the first synthetic plastics. The name came from one of the ingredients, cellulose, which makes up the cell walls in green plants and which could be obtained from sawdust or cotton. Making celluloid also required nitric acid, sulfuric acid, ether, ethyl alcohol, and camphor, combined in the right order in a tricky, dangerous procedure. A few variations in the recipe would give you not a plastic but an explosive, guncotton.

Hyatt's recipe produced a material completely different from any of its ingredients. Celluloid was hard, waterproof, and capable of being colored, molded, sawed, drilled, sanded, and polished. But it did have its drawbacks: its colors faded in sunlight; above 140 degrees Fahrenheit, it would decompose into a reddish vapor; worst of all, it was extremely flammable, especially in thin sheets. Nevertheless, Hyatt used celluloid to make billiard balls, false teeth, collars, cuffs, knobs, and handles. Celluloid had its heyday in the brief period when it was used for motion-picture film. Numerous movie-theater fires led to adoption of the safety film we use today.

Celluloid's flammability may lie behind an anecdote told by Hyatt himself in 1914. It seems that a Colorado saloonkeeper bought some celluloid-coated billiard balls. Occasionally, collisions between the cheap imitation billiard balls would make a crack loud enough to cause every man in the establishment to draw his gun.

Robert Friedel, *Pioneer Plastic: The Making and Selling of Celluloid* (Madison: University of Wisconsin Press, 1983), p. 35.

"Celluloid," *Nature,* August 19, 1880, pp. 370–371.

"Plastics," in *Encyclopaedia Britannica,* 14th ed. (1968).

Freezing Hot Water Pipes

If you have uninsulated water pipes running through a crawlspace under the house, the hot water pipe is likely to freeze before the cold water pipe during cold winter nights.

It doesn't seem to make sense. The cold water pipe is originally closer to the temperature of the outside air, so it seems logical that it would be the first to get cold enough to freeze. The monkey wrench in this argument is that hot water and cold water do not have the same freezing point—that is, the temperature required to make ice is generally lower if you start with cold water than if you start with hot water. The reason for the difference in freezing point is that cold water can hold more dissolved air than hot water. When you dissolve anything in water, you lower its freezing point.

Freezing temperature depends not only on what a substance is, but on what's dissolved in it. That's why antifreeze, dissolved in the water in a car radiator, enables that water to stay liquid below 32 degrees Fahrenheit, the freezing point of pure water. Salt does the same thing to water on the street.

Anyway, hot water holds less dissolved air than cold water, so it doesn't have to be made quite as cold before it freezes. So the hot water pipe is likely to freeze before the cold water pipe because hot water generally has less dissolved air and therefore has a higher freezing point than cold water.

This relationship between temperature and dissolved gas is also part of the reason that warm champagne and warm soda pop go flat so fast: the warm beverage loses its dissolved carbon dioxide faster than a cold beverage. This principle also explains why life is usually more abundant in cold streams than in warm streams: cold water holds more dissolved oxygen.

Ronald A. Delorenzo, *Problem Solving in General Chemistry* (Lexington, Mass.: D. C. Heath, 1981).

THE RISING WATER MYSTERY

You may well have seen this experiment in school. But did you get the right explanation? An article in the journal *The Physics Teacher* entitled "Questionable Physics Tricks for Children" points out that some of us may have gotten the wrong story.

The experiment is simple: put a burning candle in the middle of a saucer. Fill the saucer with water. Now put a drinking glass upside down over the candle, so the rim of the glass is in the water.

After a few seconds, of course, the flame uses up the oxygen in the glass and goes out. This we all expect. Now we're asked to look at the water level inside the glass. We notice that the water has risen a fraction of an inch in the glass, indicating that the volume of gas inside is less than before. And, traditionally, we're told that this shows how the oxygen in the glass has been consumed.

There are a couple of things wrong here. First, oxygen is about 20 percent of fresh air, and the volume of gas in the glass doesn't decrease by nearly that much. We have to remember that even though the flame uses up oxygen, it also makes a roughly equal volume of carbon monoxide and carbon dioxide. The same thing happens when wood burns, or any fossil fuel such as coal or oil. Second, notice how the water rises in the glass quickly, not gradually, and that it rises only after the candle goes out.

160

Now you can see what really happens when you invert a drinking glass over a candle in a saucer full of water: heat from the burning candle makes the air under the glass expand; after the candle goes out, the gas quickly cools and contracts. Pressure inside the glass therefore falls, and atmospheric pressure pushes more water up into the glass.

G. Grimvall, "Questionable Physics Tricks for Children," *The Physics Teacher,* September 1987, pp. 378–379.

WOULD YOU DRINK THIS?

Would you drink a mixture of the following ingredients: acetaldehyde, a close chemical relative of the embalming fluid formaldehyde; ethyl acetate, best known as a varnish solvent; acetone, famous as nail-polish remover; acetic acid, also known as vinegar; and a few of the compounds known as hexenals, which give freshly cut grass its characteristic odor?

It sounds horrible. But, in fact, just about all of us have drunk this mixture. These are some of the ingredients of natural grape juice. That list leaves out the three ingredients present in the greatest amounts, namely, water, sugars, and citric acid. But it's the acetaldehyde, ethyl acetate, acetone, acetic acid, and hexenals, among other substances—in small quantities and in the right proportions—that give natural grape juice its characteristic flavor.

Small quantities and correct proportions are important. Take another chemical, hydrogen cyanide, for example. Hydrogen cyanide is naturally present in small amounts in cherries, and contributes to their characteristic aroma. But in large quantities, it is a poison.

Another thing you can see from these examples is that the scientific name of a chemical is not likely to tell you whether that chemical is, so to speak, friend or foe.

Which would you rather smell: hydroxyphenol-2-butanone, or trimethylamine? You'd never know from the names alone that the first chemical contributes to the aroma of ripe raspberries, and the second causes the stench of rotten fish!

P. W. Atkins, *Molecules* (New York: Scientific American, 1987).

IS IT EVER TOO COLD TO SNOW?

Meteorologists who have looked at the records tell us that it's never too cold to snow. Snowfall has been observed at temperatures as low as 53 degrees below zero Fahrenheit.

There are several reasons we might not associate snowfall with very cold weather. First, cold air holds less water vapor than warm air. Snowflakes ultimately need water vapor to form; therefore, the less vapor, the fewer flakes. Nevertheless, there is always at least some water vapor available, no matter how cold the air gets. And under the right conditions, that vapor might produce clouds and snow.

Another reason we tend not to associate snowfall with very cold weather is that, here in the United States at least, the coldest weather tends to occur in strong high-pressure areas without clouds. In particular, we may tend to associate the deepest cold with crystal-clear nights. Clear nights feel especially cold partly because our body heat is lost directly to interplanetary space. Every warm object, including the surface of the Earth and every human being, releases heat energy in the form of invisible infrared light. On a clear night, that infrared light goes off into space, never to return. Clouds, if they're present, reflect some of that infrared light back to the ground and to us. When the infrared light hits our skin, some of its energy is converted back to heat, which makes us feel a little warmer than we would under a clear sky.

So it's never too cold to snow. But snow and extreme cold don't usually go together because the coldest nights are usually clear nights.

Donald Ahrens, *Meteorology Today*, 3rd ed. (St. Paul: West, 1987).

IS IT EVER TOO WARM TO SNOW?

Snowflakes are ice crystals; water turns to ice at a temperature no higher than 32 degrees Fahrenheit, or 0 degrees Celsius. So at first it might seem that if the temperature outside is above 32 degrees Fahrenheit, it won't snow.

Remember, though, that the atmosphere generally gets colder as you go higher. Even on a warm summer day, the air temperature above about

162

fifteen thousand feet is below 32 degrees, making it possible for snowflakes to form at high altitudes. In fact, meteorologists have concluded that most of the raindrops that fall in our part of the world originate as ice crystals which then melt as they pass through warm air on their way to the ground.

So snowflakes can form high above the ground even if the temperature near the ground is above freezing. Whether we on the ground see rain or snow depends on whether the snowflakes melt completely before they get to us.

A snowflake that has partially melted may be saved by evaporation. A snowflake in warm air picks up heat from that air. But water evaporating from the surface of a melting snowflake carries heat away from the snowflake. Energy in the form of heat is consumed in driving water molecules from the wet snowflake into the air.

Evaporation of sweat from our skin cools our bodies in exactly the same way.

Snowfall is possible at ground temperatures above 32 degrees. If evaporation can carry heat away from a falling snowflake fast enough to balance the heat coming in from warm surrounding air, that snowflake may survive long enough for us on the ground to see it.

Donald Ahrens, *Meteorology Today,* 3rd ed. (St. Paul: West, 1987).

Alcohol Content

According to the old story, a gunfighter walks into a saloon and orders a drink at the bar. He takes a shell from his gunbelt, breaks it open, and dumps a little pile of gunpowder on the counter. Then he pours some of the booze onto the gunpowder and touches a lighted match to the damp powder. If the powder burns slowly and evenly, the customer has *proof* that there's a proper amount of alcohol in the drink.

Nowadays, we look for a proof number, which is roughly twice the alcohol content: 100 proof means about 50 percent alcohol. Traditionally, of course, the alcohol in wine, beer, and liquor comes from fermentation: yeast consumes sugars or starches and makes ethyl alcohol, or ethanol, as a waste product. Sugars may come from grapes or other sweet fruits; starches from barley, wheat, corn, or potatoes, among other things.

Ethyl alcohol is poison even to the yeasts that make it. Once the alcohol content of a fermenting liquid reaches about 15 percent, the yeast dies in its own waste. That's why you don't see natural wines with alcohol content above about 15 percent.

Liquors with higher alcohol content are usually made by distillation, a process based on the fact that alcohol boils at a lower temperature than water. As a result, the vapor above a boiling liquor has a higher alcohol content than the liquor. That vapor can be passed over a cold surface, where it condenses into a new liquor, just as water vapor condenses on a cold glass. That's what happens in the coiled pipe on a moonshine still.

The distilled liquor has a higher alcohol content than the original. It also has a higher concentration of ingredients that give flavor and aroma. And, if it's not made just right, the distilled liquor will also have a dangerous concentration of potent poisons collectively named fusel oil, from the German word for "rotgut."

P. W. Atkins, *Molecules* (New York: Scientific American, 1987).
Harold McGee, *On Food and Cooking: The Science and Lore of the Kitchen* (New York: Macmillan, 1985).

Summer Antifreeze and Candy

Why use antifreeze in the summer?

Add antifreeze to the water in your car's cooling system, and you protect the cooling system from freezing in winter. The solution of antifreeze in water freezes at a much lower temperature than either water or antifreeze alone.

A general chemical principle is that when you dissolve anything in water, you lower the freezing temperature—and raise the boiling temperature of the solution. In summer heat, the coolant has to remain liquid to carry heat out of your engine. A solution of antifreeze in water protects your cooling system in summer because the solution can get hotter without boiling than can water alone.

Another application of this principle is in candy making, when you dissolve sugar in water and heat the solution. Candy recipes often specify a particular temperature to which the sugar solution is to be heated, so the sugar will crystallize in the right way and make candy of the right consistency. As you heat the water solution, water begins to boil away. The sugar stays behind, so the sugar solution becomes more concentrated.

Now you exploit the connection between concentration of sugar and boiling temperature: a more concentrated solution has a higher boiling temperature. As more water boils away, the sugar solution becomes more concentrated and the temperature of the solution rises.

Just as antifreeze raises the boiling temperature of water in your radiator, sugar raises the boiling temperature of water in a saucepan on your stove. The gradual boiling away of the water gives you precise control over the temperature of the candy mixture.

The concentration of sugar determines the temperature at which the mixture boils; the temperature at which the mixture has been boiled, in turn, determines the consistency the candy will have when it cools. The main difference between syrup, soft candy, and hard candy is in their boiling temperatures.

Ronald Delorenzo, *Problem Solving in General Chemistry* (Lexington, Mass.: D. C. Heath, 1981).

Harold McGee, *On Food and Cooking: The Science and Lore of the Kitchen* (New York: Macmillan, 1985).

Don Glass

Special Projects Director at public radio
station WFIU-FM and the radio
producer of "A Moment
of Science."

Paul Singh

Professor of Physics at Indiana University and
the science producer of "A Moment
of Science" until his retirement
in 1992.

Stephen Fentress

A writer who was most recently producer/music director at the
Strasenburgh Planetarium of the Rochester (N.Y.)
Museum and Science Center, where he wrote
and produced planetarium shows.